3rd International Conference on Plant Synthetic Biology, Bioengineering & Biotechnology 2019

Cambridge, United Kingdom
4 – 6 October 2019

ISBN: 978-1-7138-0576-2

-Printed from e-media with permission by:

Curran Associates, Inc.
57 Morehouse Lane
Red Hook, NY 12571

AIChE
120 Wall Street, FL 23
New York, NY 10005-4020

Phone: (800) 242-4363
Fax: (203) 775-5177

www.aiche.org

Additional copies of this publication are available from:

Curran Associates, Inc.
57 Morehouse Lane
Red Hook, NY 12571 USA
Phone: 845-758-0400
Fax: 845-758-2633
Email: curran@proceedings.com
Web: www.proceedings.com

TABLE OF CONTENTS

NEW TOOLS AND TECHNOLOGIES IN SYNTHETIC BIOLOGY

NOVEL BIOPRODUCTS

ENGINEERING PRIMARY PRODUCTIVITY

POSTER SESSION

Integrating Synthetic and Developmental Biology

- Conference: International Conference on Plant Synthetic Biology and Bioengineering
- Year: 2019
- Proceeding: 3rd International Conference on Plant Synthetic Biology, Bioengineering, and Biotechnology
- Group: General Submissions
- Session:
  Engineering Plant Development and Signaling
- Time:
  Friday, October 4, 2019 - 10:15am-10:40am

Authors:
Nemhauser, J., University of Washington

Life in the Anthropocene has many challenges, among the most urgent is widespread global hunger and malnourishment. What if every community of small-hold farmers could improve their own heritage crops to their own specifications? Realizing this DIY vision would require radical innovations in the way we conceptualize and execute crop improvement, as well as in how we build communities of practice. Synthetic biology, sitting at an intersection of engineering and biology, offers theoretical and practical tools to guide such efforts. My group is interested in exploring the limits of rational re-tuning of developmental pathways to customize plant form for local agricultural conditions, while simultaneously striving to understand the molecular tuning knobs that control fundamental components of eukaryotic signaling (e.g., ubiquitin-mediated protein degradation, transcriptional repression/activation).

Establishing a Platform for Genome Editing in Fresh Produce for Consumer Traits

- Conference: International Conference on Plant Synthetic Biology and Bioengineering
- Year: 2019
- Proceeding: 3rd International Conference on Plant Synthetic Biology, Bioengineering, and Biotechnology
- Group: General Submissions
- Session:
  Food and Nutritional Security
- Time:
  Friday, October 4, 2019 - 1:00pm-1:15pm

Authors:
Graham, N., Pairwise

Consumption of fruits and vegetables – the most effective solution to malnutrition – can be limited by low production or high cost in some countries, and by poor quality characteristics or inconvenient preparation in others. Americans, for example, consume only half the recommended daily intake of fruits and vegetables despite widespread availability. Pairwise seeks to address these challenges regionally and globally by making healthy food more convenient, affordable, and sustainable. To fulfill this mission, we have assembled a world class portfolio of CRISPR technologies and a state-of-the-art R&D program focused on delivering differentiated products to the fresh produce aisle. We have established base editing and cutting tools with industry-leading performance profiles in plants, including editing efficiencies as high as 85%, and extensive target sequence accessibility through gene and protein engineering. We are applying these editing tools for novel consumer traits in cane berries.

Understanding, Prospecting and Engineering Photorespiration for Increased Crop Productivity

PLANT SYNTHETIC
BIOLOGY, BIOENGINEERING,
& BIOTECHNOLOGY

- Conference: International Conference on Plant Synthetic Biology and Bioengineering
- Year: 2019
- Proceeding: 3rd International Conference on Plant Synthetic Biology, Bioengineering, and Biotechnology
- Group: General Submissions
- Session:
  Food and Nutritional Security
- Time:
  Friday, October 4, 2019 - 1:15pm-1:45pm

Authors:
South, P., USDA-ARS

Population growth, increasing global affluence and an expanding bioeconomy are conspiring to increase mid-century agricultural demand by 60–120%, a challenge that current methods of increasing crop productivity cannot meet. Traits that have increased productivity during the Green Revolution are near maximum efficiency leaving photosynthesis as the only target to double yield potential. Photorespiration is required in C3 plants to metabolize glycolate formed when Rubisco oxygenates rather than carboxylates ribulose-1,5-bisphosphate. Depending on growing temperatures photorespiration can reduce yields by 20-50% in C3 crops. Optimization of the photorespiratory process by only 5% could be worth millions annually in increased productivity. Synthetic biology has provided new opportunities in altering photorespiratory metabolism to improve photosynthetic efficiency. Using a synthetic biology approach, we can now engineer and test hundreds of prototype plants from a range of multigene construct designs aimed at changing plant metabolism. Flux through tested synthetic pathways was maximized by inhibiting glycolate export from the chloroplast. Pathways tested for alternative photorespiration improved quantum yield of photosynthesis by 20% and increased productivity by 19-37% in replicated field experiments. In addition, synthetic biology and trait screening approaches can provide insight into photosynthetic and photorespiratory metabolism providing a novel parts prospecting approach to discovering crop improvement traits. The increased efficiency identified from alternative photorespiration leads to increases in thermal tolerance and has the potential for increased food production in related species. Synthetic engineering of alternative plant metabolisms along with identifying roles of photorespiration regulation and response to abiotic stress can create a platform for engineering increased agricultural production.

Engineering Photosynthetic Electron Transfer for Increased Partitioning of Reducing Power to Cytochrome P450-Dependent Reactions

- Conference: International Conference on Plant Synthetic Biology and Bioengineering
- Year: 2019
- Proceeding: 3rd International Conference on Plant Synthetic Biology, Bioengineering, and Biotechnology
- Group: General Submissions
- Session:
  Photosynthetic Factories
- Time:
  Friday, October 4, 2019 - 3:00pm-3:15pm

Authors:
Busck Mellor, S., University of Copenhagen
Hamborg Vinde, M., University of Queensland
Nielsen, A. Z., University of Copenhagen
Thomas Hanke, G., Queen Mary University of London
Roessler, M., Imperial College
Burow, M., University of Copenhagen
Motawia, M. S., University of Copenhagen
Møller, B. L., University of Copenhagen
Jensen, P. E., University of Copenhagen

Photosynthetic organisms are promising metabolic engineering chassis. Photosynthesis allows high resource use efficiencies, and many plants can be cultivated cheaply at industrial scale. Therefore, engineering these hosts is of great relevance for future sustainable biomolecule production. Heterologous P450s can be targeted to plant thylakoid membranes, where the photosynthetic apparatus functionally substitutes native reductases via the endogenous electron carrier ferredoxin. But because ferredoxin acts as a distribution hub for photosynthetic reducing power, relying on this heavily contested source of reductant limits the photosynthesis-dependent P450 output that can be achieved. To overcome this limitation, we investigated the ability of alternative electron carriers – three plant ferredoxins and a flavodoxin-like protein - to increase photosynthesis-driven P450 activity. We also tested the electron carriers *in vivo* by fusing them to the P450 and transiently expressing these fusions targeted to chloroplasts in tobacco. Steady state kinetics reveals only modest differences in electron transfer rates between the carriers and CYP79A1 *in vitro*. However, the flavodoxin-like carrier plant P450 reductase outperforms the three ferredoxins when FNR is present to compete for reduced ferredoxin, likely because its FMN redox potential renders delivery of electrons to endogenous enzymes unfavourable. *In vivo* investigation via transient expression in tobacco show three of four CYP79A1-carrier fusions able to sequester more photosynthetic reducing power than the unfused P450. Notably fusion with the flavodoxin-like carrier yields 25-fold higher production on a per protein basis. This study demonstrates that non-native electron carriers can be used to construct novel electron transfer chains that alleviate the problem of endogenous competition. The effectiveness of such chains is governed by hey factors such as redox potential and carrier-photosystem I affinity. This demonstrates that electron transfer chain engineering and P450-carrier fusion can boosts photosynthesis-driven P450 activity and sets out concrete design parameters for future electron carrier engineering.

Introducing an Algal Pyrenoid into Higher Plants to Enhance CO2 Assimilation

PLANT SYNTHETIC BIOLOGY, BIOENGINEERING, & BIOTECHNOLOGY

- Conference: International Conference on Plant Synthetic Biology and Bioengineering
- Year: 2019
- Proceeding: 3rd International Conference on Plant Synthetic Biology, Bioengineering, and Biotechnology
- Group: General Submissions
- Session:
  Photosynthetic Factories
- Time:
  Friday, October 4, 2019 - 3:45pm-4:15pm

Authors:
McCormick, A., University of Edinburgh

Meeting the food demands of the rising global population is one of the most pressing challenges for Plant Science. Much research is now focused on overcoming the limitations of photosynthesis to improve plant productivity. One promising strategy lies in introducing a chloroplastic CO_2-concentrating mechanism (CCM) to improve the delivery of CO_2 to Rubisco, and thus increase CO_2 uptake efficiency. Eukaryotic microalgae, such as *Chlamydomonas reinhardtii*, possess highly efficient CCMs where Rubisco is aggregated in a liquid-liquid phase separated compartment called the pyrenoid. Recently we have shown that aggregation is triggered by binding of the small subunit of Rubisco to an intrinsically disordered protein called EPYC1. *In vitro* studies showed that EPYC1 can phase separate with Rubisco from a higher plant if Rubisco carries the Chlamydomonas small subunit. Here I will discuss recent work to phase-separate EPYC1 and Rubisco in a plant chloroplast and our ongoing efforts to engineer a functional algal CCM.

Synthetic Virus-like Particles and How to Make Them in Plants

PLANT SYNTHETIC BIOLOGY, BIOENGINEERING, & BIOTECHNOLOGY

- Conference: International Conference on Plant Synthetic Biology and Bioengineering
- Year: 2019
- Proceeding: 3rd International Conference on Plant Synthetic Biology, Bioengineering, and Biotechnology
- Group: General Submissions
- Session:
  Photosynthetic Factories
- Time:
  Friday, October 4, 2019 - 3:15pm-3:45pm

Authors:
Lomonossoff, G. P., John Innes Centre

Viral capsids play an essential role in the replication cycle of viruses. Not only do they protect the labile genetic material from degradation but they play a vital role in the transmission of infections both within and between host organisms. This occurs through interactions with cellular components, such as receptors, and, in many cases, with the vectors responsible for virus transmission. Furthermore, it is usually the capsid protein to which a host initially reacts when mounting a defence response, be it through adaptive or innate immunity. For this reason many vaccines are based either on inactivated virus particles or on recombinant virus-like particles (VLPs); the ability of viral capsids to elicit an immune response is also exploited to produce antibody-based diagnostic reagents. VLPs have also been exploited in bionanotechnology for such purposes as bioimaging and drug delivery. Given the importance of viral capsids in so many contexts, there has been huge interest in understanding their detailed atomic structures. Such studies have historically depended on isolating the virus of interest, finding a means of propagating it safely and efficiently in at least milligram quantities followed crystallisation and structure determination X-ray crystallography. This is a laborious process which, for many economically significant viruses, has not proved possible. However, recent developments in next-generation sequencing, bioinformatics, plant-based transient expression and cryo-electron microscopy make it possible to produce a virus capsid and determine its structure without ever needing to propagate the parent virus. To demonstrate the feasibility of this approach we have expressed the coat proteins of several viruses of both plants and animals, in plants, isolated the VLPs and determined their structures by cryo-EM. Furthermore, we have used the VLPs to develop novel diagnostic reagents and candidate vaccines.

Engineering New Rhizosphere Signalling Circuits in Cereals

PLANT SYNTHETIC BIOLOGY, BIOENGINEERING, & BIOTECHNOLOGY

- Conference: International Conference on Plant Synthetic Biology and Bioengineering
- Year: 2019
- Proceeding: 3rd International Conference on Plant Synthetic Biology, Bioengineering, and Biotechnology
- Group: General Submissions
- Session:
  Engineering Biotic Interactions
- Time:
  Saturday, October 5, 2019 - 3:00pm-3:30pm

Authors:
Paramasivan, P., Sainsbury laboratory
Geddes, B., University of Oxford
Poole, P., University of Oxford
Oldroyd, G. E. D., Sainsbury laboratory

Rhizosphere microbiome significantly influence host plant health and fitness in natural environments. Host plants reported to exude approximately 20% of their photosynthesized carbon as simple organic molecules in roots vicinity to recruit and shape its rhizosphere communities from the reservoir of microbes in soil. We aim to engineer synthetic plant to microbe signalling circuits in rhizosphere to control gene expression in rhizobacteria for delivering beneficial traits to host plants.

Rhizopines are aminocyclitol compounds produced by a few rhizobial strains in legume nodules and secreted into rhizosphere. By activating its own catabolism genes in free-living rhizobia, rhizopines provide a nutritional advantage to them during competition with other saprophytic microorganisms in rhizosphere.

I will be presenting our work on engineering rhizopine biosynthetic pathway into cereals and demonstrate the cross-kingdom signalling established between rhizopine producing transgenic plants and engineered bacteria (Geddes et. al 2019). Rhizopine mediated synthetic *trans*-kingdom signalling will enable host cereal plants to control diverse metabolism in root associated bacteria. Future work in our lab focus on using rhizopine signalling to control nitrogen fixation, phosphate solubilisation and anti-fungal metabolite production in rhizobacteria and deliver useful functions for plants.

1. Geddes B, Paramasivan P, Joffrin A, et. al. (2019) Nat Commun. 10:3430

2. Geddes et. al. (2014) Curr. Opin. Biotechnol. 32: 216-222.

UK-BBSRC and US-NSF joint funding support this work.

The Interactions between Plants and Root Microbiome in Arabidopsis and Rice

PLANT SYNTHETIC
BIOLOGY, BIOENGINEERING,
& BIOTECHNOLOGY

- Conference: International Conference on Plant Synthetic Biology and Bioengineering
- Year: 2019
- Proceeding: 3rd International Conference on Plant Synthetic Biology, Bioengineering, and Biotechnology
- Group: General Submissions
- Session:
  Engineering Biotic Interactions
- Time:
  Saturday, October 5, 2019 - 3:30pm-4:00pm

Authors:
Bai, Y., Institute of Genetics and Developmental Biology, CAS

The evolution of multiple plant traits is linked to the diverse microbe-rich soil environment. Numerous microorganisms participate in the biogeochemical transformation of soil nutrients, including nitrogen and phosphorus cycling, which are essential for plant productivity. Plant roots selectively recruit defined root microbiota from soil in both model and crop plants. It has been reported that root microbiota are determined by the rhizocompartment, soil type, geographical location, nutrient status, developmental stage and host genotype. However, variation in the root microbiota among a plant population and the linked plant genetic basis are not yet clear. We use *Arabidopsis* and rice as models, and found that triterpene metabolites selectively modulate root microbiome members and plants coordinate root microbiome to utilize soil nitrogen.

Transcriptional Rewiring: Engineering Plant Stress Response

PLANT SYNTHETIC
BIOLOGY, BIOENGINEERING,
& BIOTECHNOLOGY

- Conference: International Conference on Plant Synthetic Biology and Bioengineering
- Year: 2019
- Proceeding: 3rd International Conference on Plant Synthetic Biology, Bioengineering, and Biotechnology
- Group: General Submissions
- Session:
  Engineering Biotic Interactions
- Time:
  Saturday, October 5, 2019 - 4:00pm-4:30pm

Authors:
Windram, O., Imperial College London

Plant defence responses are modulated by substantial transcriptional reprogramming, up to 40% of the genome can be differentially expressed following pathogen challenge. These differentially expressed genes include hundreds of transcription factors suggesting a complex transcriptional network regulates plant stress responses. I will present how we are using transcriptional rewiring, of plant defence regulators, to engineer plants with enhanced tolerance to pathogens. Transcriptional rewiring involves the fusion of transcription factor coding sequences with non-natural promoter sequences, which, in turn can have pleotropic cascade effects on downstream target pathways. We have previously shown that transcriptional rewiring can be used to generate novel phenotypic diversity that can adapt yeast to stresses associated with heterologous expression. Here I will show how we are using a combination of Synthetic and Systems Biology, inspired by our studies in yeast, to select plant regulators for targeted rewiring approaches to enhance plant tolerance to pathogens. First, we model the transcriptional network using transcriptome time series data. From this network we select genes with characteristic network topological properties. Promoters and coding sequences of the selected genes are then rewired and constructs screen for altered stress response. Rewired regulators exhibit a range of altered plant stress response phenotypes highlighting the ability of our modelling to identify key network components of stress response. Importantly, several rewired constructs appear to improve tolerance to stress suggesting such approaches could be used to design new crops with improved stress tolerance.

Genome Editing with Programmable Nucleases in Crop Plants

PLANT SYNTHETIC
BIOLOGY, BIOENGINEERING,
& BIOTECHNOLOGY

- Conference: International Conference on Plant Synthetic Biology and Bioengineering
- Year: 2019
- Proceeding: 3rd International Conference on Plant Synthetic Biology, Bioengineering, and Biotechnology
- Group: General Submissions
- Session:
  Keynote 3
- Time:
  Saturday, October 5, 2019 - 8:45am-9:30am

Authors:
Gao, C., Institute of Genetics and Developmental Biology

Crop improvement requires the constant creation and use of new allelic variants. Conventional breeding can be limited in providing the genes and alleles required to meet the agricultural challenges. In the past decade, Genome editing can accelerate plant breeding by allowing the introduction of precise and predictable modifications directly in an elite background. The most promising utilization of the CRISPR/Cas9 system can be used to generate targeted genome modifications including mutations, insertions, replacements and chromosome rearrangements. The use of CRISPR in agriculture should be considered as simply a new breeding method that can produce identical results to conventional methods in a much more predictable, faster and even cheaper manner.

Computational Tools for the Discovery of Novel Specialized Metabolic Pathways in Plants

PLANT SYNTHETIC
BIOLOGY, BIOENGINEERING,
& BIOTECHNOLOGY

- Conference: International Conference on Plant Synthetic Biology and Bioengineering
- Year: 2019
- Proceeding: 3rd International Conference on Plant Synthetic Biology, Bioengineering, and Biotechnology
- Group: General Submissions
- Session:
  New Tools and Technologies in Synthetic Biology
- Time:
  Saturday, October 5, 2019 - 11:30am-12:00pm

Authors:
Medema, M., Wageningen University

Plants produce a wide variety of secondary metabolites: chemical compounds that play key roles as toxins in their defence against pathogens and insects, as scents and pigments to attract pollinators and as signalling molecules to communicate with other plants, animals and microbes. However, only very few of their biosynthetic pathways have been identified, and the mechanisms underpinning metabolic diversification in evolution are largely unknown. The recent accumulation of genomic, transcriptomic and metabolomic data provides new opportunities. Here, I will describe computational tools that leverage diverse types of omics data to identify plant biosynthetic pathways and analyze their diversity and evolution, and outline the prospects of leveraging the results for pathway engineering in synthetic biology.

CRISPR-Enabled Complex Trait Loci for Gene Stacking in Maize

PLANT SYNTHETIC BIOLOGY, BIOENGINEERING, & BIOTECHNOLOGY

- Conference: International Conference on Plant Synthetic Biology and Bioengineering
- Year: 2019
- Proceeding: 3rd International Conference on Plant Synthetic Biology, Bioengineering, and Biotechnology
- Group: General Submissions
- Session:
  New Tools and Technologies in Synthetic Biology
- Time:
  Saturday, October 5, 2019 - 11:15am-11:30am

Authors:
Betts, S. D., Corteva Agriscience
Gao, H., Corteva Agriscience
Young, J. K., Corteva Agriscience
Mutti, J., Corteva Agriscience
Yang, M., Corteva Agriscience
Schroder, M., Corteva Agriscience
Lenderts, B., Corteva Agriscience
Wang, L., Corteva Agriscience
St. Claire, G., Corteva Agriscience
Jones, S., Corteva Agriscience
Feigenbutz, L., Corteva Agriscience
Zeng, M., Corteva Agriscience
Wagner, S., Corteva Agriscience
Sopko, X., Corteva Agriscience
Sander, J. D., Corteva Agriscience
Chilcoat, N. D., Corteva Agriscience

Sustainable maize production requires combining multiple biotech and native traits that provide insect control, disease resistance and tolerance to herbicides or abiotic stresses. Because trait genes have been randomly inserted at various genomic locations, combining numerous traits is expensive, time-consuming and complex. Here we report the Complex Trait Locus (CTL) for genetic trait stacking. CTL consisted of multiple preselected sites which were positioned within a small, well characterized chromosomal region. Firstly, a site-specific insertion landing pad (SSILP) was targeted to one of the sites using CRISPR-Cas9 technology, producing individual inbred lines capable of receiving trait genes via recombinase-mediated cassette exchange (RMCE). The traits then can be combined within a linkage block using genetic crossing and bred as a single locus. We generated 57 SSILP lines at 4 CTLs across 3 inbreds. The *NPTII* gene in the SSILP expressed consistently and had minimal impact on neighboring endogenous gene expression. Two-gene stacking was readily achieved. The CTL technology is a major step forward in streamlining biotech product testing and deployment. This study also advances our understanding of gene targeting and expression.

Designing Chloroplast Expression Systems with Predictable Protein Output

PLANT SYNTHETIC
BIOLOGY, BIOENGINEERING,
& BIOTECHNOLOGY

- Conference: International Conference on Plant Synthetic Biology and Bioengineering
- Year: 2019
- Proceeding: 3rd International Conference on Plant Synthetic Biology, Bioengineering, and Biotechnology
- Group: General Submissions
- Session:
  New Tools and Technologies in Synthetic Biology
- Time:
  Saturday, October 5, 2019 - 10:30am-10:45am

Authors:
Maliga, P., Rutgers University
Yu, Q., Rutgers University
Barkan, A., University of Oregon

Transgenes in leaves from native expression signals are expressed at such high levels that gene expression compromises growth, while in non-green plastids protein levels are very low. We developed a two-part regulatory system that relies on (1) translation control sequences of plastid transgene mRNAs that respond to (2) nuclear genes that selectively stabilize and facilitate translation of the plastid mRNAs. The engineered *atpH* mRNA translation control sequences are upstream of the AUG in plastid mRNAs. The nuclear genes encode engineered PPR10 proteins that recognize the cognate site. In potato, GFP from wild-type signals accumulates up to 25% of total soluble cellular protein (TSP) in leaves that seriously compromises plant development. The engineered binding site reduces GFP accumulation to levels that do not impact growth, while expression of the engineered PPR10 from a tuber-specific promoter boosts GFP accumulation 20x. Based on studies in the model plant tobacco, dependent on the tissue type, the two-part system enables regulated expression of plastid genes in the range of 0.05% to 25% of TSP. The two-part system enables regulated plastid transgene expression without compromising yield.

Looking Beyond GWAS: Identifying Functional Roles of SNPs Using Metabolic Networks in Arabidopsis and Populus

PLANT SYNTHETIC BIOLOGY, BIOENGINEERING, & BIOTECHNOLOGY

- Conference: International Conference on Plant Synthetic Biology and Bioengineering
- Year: 2019
- Proceeding: 3rd International Conference on Plant Synthetic Biology, Bioengineering, and Biotechnology
- Group: General Submissions
- Session:
  New Tools and Technologies in Synthetic Biology
- Time:
  Saturday, October 5, 2019 - 12:15pm-12:30pm

Authors:
Sarkar, D., The Pennsylvania State University
Maranas, C., The Pennsylvania State University

Studies aimed at improving agricultural yield and understanding adaptive processes have focused on explaining genetic sources of phenotypic variation. GWAS identifies the genetic background behind a trait by examining associations between phenotypes and single-nucleotide polymorphisms (SNPs). Although such studies are common, they often lack the power required to uncover relatively small effect sizes conferred by most loci and are not directly informative of the governing biological mechanisms. SNPeffect is a complementary analysis that offers genotype-to-phenotype interpretations by constructing “scenarios” for the mechanistic roles of SNPs in enzyme-coding (or regulatory) regions, integrating metabolomics, proteomics, and metabolic network information into a self-consistent narrative. SNPeffect can thus be used *a posteriori* to functionally explain GWAS hits or *a priori* to generate linkages between SNPs, genes, enzymes, metabolites and phenotypes, to be used as priors in GWAS and other plant breeding techniques.

SNPeffect was used to explain differential growth and metabolite accumulation in *A.thaliana* and *P. trichocarpa* accessions as the outcome of activating and inactivating SNPs in enzyme-coding regions of the genotypes. To this end, we also constructed a *P.trichocarpa* genome-scale metabolic model, first for a perennial woody tree. As expected, our results indicate that plant growth is a complex polygenic trait governed by carbon and energy partitioning. Growth-affecting SNPs were found to be primarily in glycolysis, amino-acid, and energy metabolism. Faster-growing *Arabidopsis* genotypes were predicted to have higher fluxes through protein metabolism pathways and employ the energy efficient purine salvage pathway as opposed to *de novo* purine biosynthesis for generating nucleotides. In poplar, growth-determining SNPs with an upregulating role were in genes from cellulose and lignin biosynthesis, amino acid, and energy metabolism, in line with breeding strategies targeting pathways governing carbon and energy partition. SNPeffect also identified (putatively) functional SNPs in both species, distributed among pathways such as glycolysis, folate, and pyrimidine metabolism.

Microbial Cell Factories for Plant Secondary Metabolites - Engineering Glucosinolates in Bacteria and Yeast

PLANT SYNTHETIC BIOLOGY, BIOENGINEERING, & BIOTECHNOLOGY

- Conference: International Conference on Plant Synthetic Biology and Bioengineering
- Year: 2019
- Proceeding: 3rd International Conference on Plant Synthetic Biology, Bioengineering, and Biotechnology
- Group: General Submissions
- Session:
  New Tools and Technologies in Synthetic Biology
- Time:
  Saturday, October 5, 2019 - 10:15am-10:30am

Authors:
Poborsky, M., University of Copenhagen
Halkier, B. A., University of Copenhagen
Crocoll, C., University of Copenhagen
Petersen, A., University of Copenhagen

Glucosinolates are amino acid-derived sulphur-rich specialised metabolites characteristic of the Brassicales order and well known for their health-promoting effects upon intake of e.g. broccoli and cabbage. However, high amounts of broccoli and cabbage must be eaten on a regular basis to fully realize the health-beneficial effects and the current supplement market is based on products with unreliable amounts, if any at all. This has primed a strong desire to develop an efficient microbial cell factory for glucosinolate production as stable, rich source.

Building upon our long lasting experience with glucosinolate pathway reconstruction and enzyme discovery, we have moved onto engineering the synthetic pathway in Escherichia coli and Saccharomyces cerevisiae. We have successfully established the production of benzyl glucosinolate (BGLS)[1] and indol-3-ylmethyl (I3M)[2] glucosinolate as well as we have made substantial advancements in the amino acid chain elongation[3,4], which is a necessary first step for some of the most potent glucosinolates. We were able to boost the synthesis of BGLS, by systematic pathway engineering, optimization of the flux through the entry point cytochrome P450 enzymes and the last sulfotransferase step.

Towards our goal of engineering microbial high-level production of glucosinolates, we have encountered challenges related to the microbial sulphur chemistry to produce the sulphur rich glucosinolates and to premature abortion of the pathway by the intermediates escaping from the host cells. We apply pathway engineering as well as transport engineering approaches to overcome these challenges. Learnt lessons will be discussed.

1. Petersen, *Metabolic engineering* (2019)
2. Mikkelsen, *Metabolic engineering* (2012)
3. Mirza, *Metabolic engineering* (2016)

4. Petersen, *Bioscience reports* (2019)

Plant Reactome Knowledgebase: A Resource for Plant Synthetic Biology

PLANT SYNTHETIC
BIOLOGY, BIOENGINEERING,
& BIOTECHNOLOGY

- Conference: International Conference on Plant Synthetic Biology and Bioengineering
- Year: 2019
- Proceeding: 3rd International Conference on Plant Synthetic Biology, Bioengineering, and Biotechnology
- Group: General Submissions
- Session:
  New Tools and Technologies in Synthetic Biology
- Time:
  Saturday, October 5, 2019 - 12:00pm-12:15pm

Authors:
Naithani, S., Oregon State University
Preece, J., Oregon State University
Gupta, P., Oregon State University
D'Eustachio, P., NYU School of Medicine
Elser, J., Oregon State University
Jaiswal, P., Oregon State University

In-silico systems-level pathway knowledgebases are key reference resources that provide conceptual frameworks for plant biologists aiming to design, model, and test synthetic metabolic pathways in plants. Plant Reactome (https://plantreactome.gramene.org) is an open-source, free-of-cost pathway portal of the Gramene project (www.gramene.org). Plant Reactome features plant metabolic, signaling, transport, genetic, regulatory, developmental and stress-response pathways for 83 plant species, from unicellular photoautotrophs to higher plants. We utilize a combinatorial approach involving manual curation of reference pathways and automated gene orthology-based projection to rapidly scale up pathway knowledge from reference species rice (*Oryza sativa*) to all other species. Pathway clustering across the broad phylogenetic spectrum of photosynthetic organisms shows distinct gene-pathway association patterns reflecting evolutionary history and ploidy levels. Plant researchers can compare reference rice pathways with the projected pathways from any hosted plant species to discover species-specific pathway enrichment, regulatory events, and loss/gain of reactions, thereby providing necessary clues and hypotheses for building potential pathway engineering and synthetic biology strategies for the improvement of the physiology, utility, performance and synthesis of bioproducts. Users can also upload their data for pathway analysis, visualization and integration of gene-gene interactions, and access Plant Reactome pathway data in various standardized formats via an embeddable widget, web services, and other download mechanisms. The Plant Reactome is funded by the NSF (IOS-1127112), the NIH (U41 HG003751), ENFIN (LSHG-CT-2005-518254), the Ontario Research Fund, and the EBI Industry Programme.

Rapidly Reprogramming Plants with RNA Vectors

PLANT SYNTHETIC
BIOLOGY, BIOENGINEERING,
& BIOTECHNOLOGY

- Conference: International Conference on Plant Synthetic Biology and Bioengineering
- Year: 2019
- Proceeding: 3rd International Conference on Plant Synthetic Biology, Bioengineering, and Biotechnology
- Group: General Submissions
- Session:
  New Tools and Technologies in Synthetic Biology
- Time:
  Saturday, October 5, 2019 - 11:00am-11:15am

Authors:
Khakhar, A., University of Minnesota
Voytas, D., University of Minnesota
Smanski, M. J., University of Minnesota, Twin Cities

Transcriptional programs sculpt plant morphology and metabolism to meet environmental challenges. These same programs have been manipulated through selective breeding over generations to increase agricultural productivity and robustness. Studying the genetic basis of these improved traits has elucidated the specific alterations in the expression levels of key master regulator genes that lead to phenotypic improvement. The use of synthetic transcription factors has enabled precise alterations to the transcriptional landscape, but their application in crop plants has been challenging due to difficulties posed by plant transformation and long life cycles of crops. We are developing a system called VipariNama (ViN, from the Sanskrit: to change) to allow rapid reprogramming of transcription in adult plants. It uses systemically mobile and stably maintained RNA scaffolds in conjunction with genomically integrated transcription factor components to assemble transcription factors at loci of interest. To experimentally validate the ViN system, we have built lines of the model plants *Arabidopsis thaliana, Nicotiana benthamiana* and the crop *Oryza sativa* with genomically integrated copies of functionally validated Cas9-based activators and repressors. We demonstrate the capacity to alter gene expression via guide RNAs delivered on mobile RNA viruses. We also explore how these transcription factors themselves can be delivered via viral ViN vectors. Finally, we are optimizing the systemic mobility and stability of our RNA vectors by using next generation sequencing enabled screening of variant libraries. By putting these optimized ViN vectors and deconstructed transcription factors together we hope to facilitate rapid reprogramming of transcriptional landscapes in adult plants and thereby accelerate the phenotypic improvement of crops as well as enable generation of necessary traits in the field to ensure food security in rapidly changing agricultural environments.

Synthetic Biology and Fermentation Platforms for the Production of High-Value Plant Natural Products

PLANT SYNTHETIC BIOLOGY, BIOENGINEERING, & BIOTECHNOLOGY

- Conference: International Conference on Plant Synthetic Biology and Bioengineering
- Year: 2019
- Proceeding: 3rd International Conference on Plant Synthetic Biology, Bioengineering, and Biotechnology
- Group: General Submissions
- Session:
  New Tools and Technologies in Synthetic Biology
- Time:
  Saturday, October 5, 2019 - 10:00am-10:15am

Authors:
Lu, W., Conagen Inc.

Many wonderful molecules found in nature are scarce, impure, or hard to consistently supply. In particular, many fragrances and flavors are derived from plants. Extraction of these ingredients has its downsides, from environmental hazards to burdens on our natural resources to production inefficiencies. At Conagen Inc., we have established powerful synthetic biology and fermentation platforms to produce high-quality, rare and costly ingredients. As an example, we will talk about one of our representative terpene products α-ionone, which is a derivative of carotenoids and valued in perfumery and flavor. Our Natural-Terpene Platform covers gene discovery, strain engineering, process optimization, fermentation scale-up, and downstream processing. We will share several important insights for strain development including heterologous host selection, precursor supply improvement, bottleneck identification, novel gene discovery, protein engineering, solubility enhancement, and subcellular compartmentalization. Some representative products from our Natural-Terpene Platform will also be introduced in the presentation.

Synthetic Biology and Microfluidics to Engineer Lipid Metabolism in Marine Algae

PLANT SYNTHETIC
BIOLOGY, BIOENGINEERING,
& BIOTECHNOLOGY

- Conference: International Conference on Plant Synthetic Biology and Bioengineering
- Year: 2019
- Proceeding: 3rd International Conference on Plant Synthetic Biology, Bioengineering, and Biotechnology
- Group: General Submissions
- Session:
  New Tools and Technologies in Synthetic Biology
- Time:
  Saturday, October 5, 2019 - 10:45am-11:00am

Authors:
Geisler, K., University of Cambridge
Smith, A., University of Cambridge
Yu, Z., University of Cambridge

Marine microalgae like the diatom *Phaeodactylum tricornutum* are exciting biotechnological chassis as they can grow to high density on non-potable water sources. The availability of genomic resources and the ability to accumulate naturally large amounts of lipids in form of triacylglycerols (TAGs) make *P. tricornutum* an interesting host to produce novel fatty acids or valuable oils for food and feed. Over the last few years, we and others have developed a range of established and novel DNA parts (constitutive and inducible promoters, 5'UTRs, 3'UTRs and target peptides) for *P. tricornutum* based on the MoClo syntax and Golden Gate assembly. A similar toolkit has been recently established for the green algae *Chlamydomonas reinhardtii* (Crozet et al., 2018). Using these DNA parts, we have assembled molecular devices to manipulate the lipid metabolism in *P. tricornutum*. We will present data on the overexpression of diacylglycerol acyltransferases (DGATs) and phospholipid:diacylglycerol acyltransferase (PDAT), enzymes involved in TAG synthesis. In addition, we are combining metabolic engineering with state-of-the-art microfluidic devices for the encapsulation and growth of wildtype and transformed microalgae cells in microdroplets, allowing the analysis of single cells and the screening of them in a high-throughput manner. The sorting technology can be used to screen for highly expressing lines, and to detect strains within a mutagenized population with desirable characteristics.

Ref.: Crozet P, et al (2018) Birth of a Photosynthetic Chassis: A MoClo Toolkit Enabling Synthetic Biology in the Microalga *Chlamydomonas reinhardtii*. ACS Synth Biol 7: 2074–2086

Use of mRNA Secondary Structure to Fine-Tune and Regulate Translation in Chloroplasts

PLANT SYNTHETIC BIOLOGY, BIOENGINEERING, & BIOTECHNOLOGY

- Conference: International Conference on Plant Synthetic Biology and Bioengineering
- Year: 2019
- Proceeding: 3rd International Conference on Plant Synthetic Biology, Bioengineering, and Biotechnology
- Group: General Submissions
- Session:
  New Tools and Technologies in Synthetic Biology
- Time:
  Saturday, October 5, 2019 - 12:30pm-12:45pm

Authors:
Scharff, L. B., University of Copenhagen

Chloroplasts enable the production of very high amounts of heterologous proteins, up to 70% of total soluble protein. In addition, chloroplasts harbour many metabolic pathways of interest for synthetic biology approaches. It is therefore important to understand what influences protein levels and how gene expression is regulated. A major determinant is translation. The efficiency of translation initiation depends on start codon recognition. Start codons are defined by Shine-Dalgarno-sequences and/or local minima of mRNA secondary structure. Based on *in vitro* data it was hypothesised that translation initiation is regulated by RNA binding proteins that modify the mRNA secondary structure of the translation initiation region and thereby its accessibility. We tested this hypothesis by analysing *in vivo* mRNA secondary structure using two complementary methods (SHAPE-seq and DMS-MaPseq) and by analyzing translation by ribosome profiling. The Shine-Dalgarno sequence and the start codon of *psbA*, whose translation is upregulated in high light, are more accessible in conditions, which induce translation of *psbA*. Furthermore, we detected a footprint of a regulatory protein in the 5' UTR of *psbA*. It is such located that its binding can open the structure of translation initiation region and thereby making it accessible for the ribosome. Moreover, also other genes with weak Shine-Dalgarno sequences are regulated like *psbA*, while those with strong Shine-Dalgarno sequences are not. This indicates that the modulation of the accessibility of the translation initiation region by RNA binding proteins is a general mechanism to regulate translation initiation in plastids. These results highlight the importance of optimising the mRNA secondary structure of the gene of interest to obtain the desired protein levels.

Breakthroughs in Plant Based PHB Production: Harnessing Nature to Heal Nature

PLANT SYNTHETIC BIOLOGY, BIOENGINEERING, & BIOTECHNOLOGY

- Conference: International Conference on Plant Synthetic Biology and Bioengineering
- Year: 2019
- Proceeding: 3rd International Conference on Plant Synthetic Biology, Bioengineering, and Biotechnology
- Group: General Submissions
- Session:
  Novel Bioproducts
- Time:
  Saturday, October 5, 2019 - 2:30pm-3:00pm

Authors:
Snell, K. D., Yield10 Bioscience
Malik, M. R., Metabolix Oilseeds
Peoples, O. P., Yield10 Bioscience

Crops have the potential to not only be a source of food and fuel but also a source of renewable materials. One of the most interesting classes of material targets are the polyhydroxyalkanoate (PHA) family of microbial carbon and energy storage polymers, which have a wide range of potential applications. Although PHAs are targeted as replacements for petroleum derived plastics, and over time these markets will dominate, water treatment is a simpler market entry strategy and directly linked to improving the sustainability of food production and consumption. PHAs are well known growth-substrates for denitrifying bacteria in water systems enabling them to convert nitrate to nitrogen gas and reduce nitrate levels produced from food production and human waste. Developing PHA producing oilseed cover crops such as *Camelina sativa* to reduce nitrogen runoff in agricultural fields would also enable low cost production of PHA pellets for water treatment applications. Oilseeds are an ideal production platform since seed metabolism is well suited to the polymer pathways and multiple co-products can be harvested (polymer, seed oil, protein rich seed meal) increasing the value of the seed.

Yield10 has made significant progress in addressing the technical challenges of producing polyhydroxybutyrate (PHB), the simplest member of the PHA family, in plants. We previously demonstrated the production of high levels of PHB in the seeds of Camelina by targeting enzymes for PHB synthesis to the seed plastids, however the seedlings were severely impaired in emergence and survival. Yield10 has recently demonstrated a cytosolic PHB production pathway with an ER targeted polymerizing enzyme PHA synthase that produced PHB levels up to 10.2% of the seed weight in T4 seeds with good seedling emergence and survival. These results are a significant step forward towards commercial production of PHB in plants.

Building Non-Canonical Isoprenoid Pathways in Yeast

PLANT SYNTHETIC BIOLOGY, BIOENGINEERING, & BIOTECHNOLOGY

- Conference: International Conference on Plant Synthetic Biology and Bioengineering
- Year: 2019
- Proceeding: 3rd International Conference on Plant Synthetic Biology, Bioengineering, and Biotechnology
- Group: General Submissions
- Session:
  Novel Bioproducts
- Time:
  Saturday, October 5, 2019 - 1:30pm-2:00pm

Authors:
Kampranis, S., University of Copenhagen

One application of synthetic biology is the re-design of existing biological systems to acquire new functions. In this context, expanding the chemical code underlying key biosynthetic pathways will lead to the synthesis of compounds with new structures and potentially new biological activities. Terpenoids are the largest group of specialized metabolites with numerous applications. Yet, being synthesized from 5-carbon units, they are restricted to distinct classes that differ by 5 carbon atoms (C10, C15, C20, etc.). To expand the diversity of terpenoid structures, we engineered yeast cells to synthesize a non-canonical building block with 11 carbons, and produced 40 C11 terpene scaffolds that can form the basis for an entire terpenoid class. By identifying a single-residue switch that converts C10 plant monoterpene synthases to C11-specific enzymes, we engineered dedicated synthases for C11 terpene production. This approach will enable the systematic expansion of the chemical space accessed by terpenoids.

Reference:

1. Ignea, C., Pontini, M., Motawia, M. S., Maffei, M. E., Makris, A. M. & Kampranis, S. C. (2018) Synthesis of 11-carbon terpenoids in yeast using protein and metabolic engineering, ***Nature Chemical Biology***, 14, 1090-98.

Engineering the Stereoisomeric Structure of Plant Triacylglycerol to Mimic Human Milk Fat

PLANT SYNTHETIC
BIOLOGY, BIOENGINEERING,
& BIOTECHNOLOGY

- Conference: International Conference on Plant Synthetic Biology and Bioengineering
- Year: 2019
- Proceeding: 3rd International Conference on Plant Synthetic Biology, Bioengineering, and Biotechnology
- Group: General Submissions
- Session:
  Novel Bioproducts
- Time:
  Saturday, October 5, 2019 - 2:00pm-2:30pm

Authors:
Eastmond, P., Rothamsted Research

Human milk fat substitute (HMFS) is a class of structured lipid that is widely used as an ingredient in infant formulas. Like human milk fat, HMFS is characterised by an enrichment of palmitoyl (C16:0) groups at the middle (sn-2 or β) position on the glycerol backbone, and there is evidence from multiple clinical trials that triacylglycerol (TAG) with this unusual stereoisomeric structure assists nutrient absorption in the neonatal gut. HMFS production currently relies on enzyme-based catalysis since there is no biological source of fat with the equivalent structure, other than humans. Around 1/2 a million tons of fat is used in infant formula production each year. The majority of this fat is obtained from plants, which specifically exclude C16:0 from the middle position. In this study we have engineered the metabolic pathway for TAG biosynthesis in the model oilseed *Arabidopsis thaliana* so that the percentage of C16:0 esterified to the middle (versus outer) position(s) is increased by more than 20-fold (i.e. from ~3% in wild type to >70% in our final iteration). We have also modified fatty acid biosynthesis so that total fatty acyl composition, as well as TAG structure, is more comparable to that of human milk fat. Finally, we have applied this technology to the oilseed crop *Brassica napus* and oleaginous yeast *Yarrowia lipolytica*. The work we will present therefore provides new biological sources of HMFS for infant formula.

From Discovery to Market: Fermentation for Ingredients and Natural Products

PLANT SYNTHETIC
BIOLOGY, BIOENGINEERING,
& BIOTECHNOLOGY

- Conference: International Conference on Plant Synthetic Biology and Bioengineering
- Year: 2019
- Proceeding: 3rd International Conference on Plant Synthetic Biology, Bioengineering, and Biotechnology
- Group: General Submissions
- Session:
  Novel Bioproducts
- Time:
  Saturday, October 5, 2019 - 3:30pm-4:00pm

Authors:
Chen, H., Conagen
Mattozzi, M., Conagen Inc

Plants and their secondary metabolites have served as fragrances, flavors, and pharmaceuticals for millennia. While the discovery of new secondary metabolites is extremely important, cost barriers such as limited availability and high molecule complexity can prevent their adoption by consumers. The emerging field of synthetic biology is beginning to democratize this process. By determining the metabolic pathways by which plants make small molecules and re-constructing them an organism amenable to fermentation, we can greatly increase the availability of natural products. Here we present case studies in the natural flavorings (peach lactones) and sweeteners (steviol glycosides) space and how Conagen has produced natural product molecules heterologously via fermentation. Current and future applications of bioproduction will be discussed, and several case studies will be presented that illustrate the impact already being felt in the food and beverage space. We are interested in working with academics to sponsor projects for natural product and metabolic pathway discovery.

Metabolic Engineering of Microbial Cell Factories for the Production of High Value Products from Plant Fatty Acid Metabolism

PLANT SYNTHETIC
BIOLOGY, BIOENGINEERING,
& BIOTECHNOLOGY

- Conference: International Conference on Plant Synthetic Biology and Bioengineering
- Year: 2019
- Proceeding: 3rd International Conference on Plant Synthetic Biology, Bioengineering, and Biotechnology
- Group: General Submissions
- Session:
  Novel Bioproducts
- Time:
  Saturday, October 5, 2019 - 1:00pm-1:30pm

Authors:
Chen, H., Conagen Inc
Yu, O., Conagen Inc.

Fatty acid-derived biofuels and oleochemicals are an important group of target compounds in the field of synthetic biology. These compounds encompass a wide range of fatty acid metabolism such as free fatty acids, fatty alcohols, fatty aldehydes, fatty esters, alkanes, olefins, methyl ketones and waxes. At Conagen, our product development has been focused on fatty acid-derived flavor and fragrance chemicals with high economic value. In this presentation, we will talk about some of our representative products including hydroxy fatty acid-derived γ- and δ- lactones which are fruity, creamy and milky; α- linolenic acid-derived grassy *cis*-3-hexenol as well as jasmonates; and branched medium chain length fatty acid-derived capsaicinoids, the active ingredients in hot peppers. Our Natural-Lipo Platform covers gene discovery, strain engineering, process optimization, fermentation scale-up and downstream process. In strain development, we couple the screening of natural producers with pathway engineering of conventional microbes for different product features. Some examples will be given in the presentation.

Synthesis of Non-Canonical Terpenoids in Yeast

PLANT SYNTHETIC
BIOLOGY, BIOENGINEERING,
& BIOTECHNOLOGY

- Conference: International Conference on Plant Synthetic Biology and Bioengineering
- Year: 2019
- Proceeding: 3rd International Conference on Plant Synthetic Biology, Bioengineering, and Biotechnology
- Group: General Submissions
- Session:
  Novel Bioproducts
- Time:
  Saturday, October 5, 2019 - 3:00pm-3:30pm

Authors:
Kampranis, S., University of Copenhagen

One application of synthetic biology is the re-design of existing biological systems to acquire new functions. In this context, expanding the chemical code underlying key biosynthetic pathways will lead to the synthesis of compounds with new structures and potentially new biological activities. Terpenoids are the largest group of specialized metabolites with numerous applications. Yet, being synthesized from 5-carbon units, they are restricted to distinct classes that differ by 5 carbon atoms (C10, C15, C20, etc.). To expand the diversity of terpenoid structures, we engineered yeast cells to synthesize a non-canonical building block with 11 carbons and produced 40 C11 terpene scaffolds that can form the basis for an entire terpenoid class. By identifying a single-residue switch that converts C10 plant monoterpene synthases to C11-specific enzymes, we engineered dedicated synthases for C11 terpene production. This approach will enable the systematic expansion of the chemical space accessed by terpenoids.

Reference:

Ignea, C., Pontini, M., Motawia, M. S., Maffei, M. E., Makris, A. M. & Kampranis, S. C. (2018) Synthesis of 11-carbon terpenoids in yeast using protein and metabolic engineering, ***Nature Chemical Biology***, 14, 1090-98.

Daring metabolic designs for enhanced plant carbon fixation

PLANT SYNTHETIC
BIOLOGY, BIOENGINEERING,
& BIOTECHNOLOGY

- Conference: International Conference on Plant Synthetic Biology and Bioengineering
- Year: 2019
- Proceeding: 3rd International Conference on Plant Synthetic Biology, Bioengineering, and Biotechnology
- Group: General Submissions
- Session:
  Engineering Primary Productivity
- Time:
  Sunday, October 6, 2019 - 10:30am-11:00am

Authors:
Bar-Even, A., Max Planck Institute for Molecular Plant Physiology

Increasing agricultural productivity is one of the major challenges our society faces. While multiple strategies to enhance plant carbon fixation have been suggested, and partially implemented, most of them are restricted to relatively simple modifications of endogenous metabolism, i.e., "low hanging fruit". In this talk I aim to portray the next generation of metabolic solutions to increase carbon fixation rate and yield. These strategies involve major rewiring of central metabolism, including engineering photorespiration bypass routes that do not release carbon and supporting the Calvin Cycle by establishing a parallel carbon fixation route based on CO_2 reduction to formate. While the barriers for implementing these elaborated metabolic architectures are quite significant, if we truly want to revolutionize carbon fixation, only daring engineering efforts will lead the way.

Functional Reconstitution of the Bacterial CO2 Concentrating Mechanism

PLANT SYNTHETIC
BIOLOGY, BIOENGINEERING,
& BIOTECHNOLOGY

- Conference: International Conference on Plant Synthetic Biology and Bioengineering
- Year: 2019
- Proceeding: 3rd International Conference on Plant Synthetic Biology, Bioengineering, and Biotechnology
- Group: General Submissions
- Session:
  Engineering Primary Productivity
- Time:
  Sunday, October 6, 2019 - 11:00am-11:15am

Authors:
Savage, D., UC Berkeley

Bacterial autotrophs often rely on biophysical CO_2 concentrating mechanisms (CCMs) to assimilate carbon. It is postulated that principles - and perhaps even components – of these CCMs could be used to improve CO_2 assimilation in plants. Surprisingly, a systematic 'parts list' of the CCM remains unknown. To this end, we have carried out a genome-wide barcoded transposon screen to identify essential and CCM-related genes in the ɣ-proteobacterium *H. neapolitanus*. Screening revealed that the CCM comprises at least 17 and likely no more than 25 genes, most of which are encoded in 3 operons. Found within these genes is a new class of C_i pump which we term DAB, for 'DABs accumulate bicarbonate,' that is widespread amongst prokaryotes and readily expressed in a heterologous fashion. Informed by this information, we have also performed a functional reconstitution of the bacterial CCM *in vivo*. Using a novel strain of Rubisco-dependent *E. coli*, we have assessed the importance of known and poorly characterized activities of the CCM for achieving efficient CO_2 assimilation and, ultimately, have engineered a strain which is capable of growth due to efficient CO_2 fixation directly from ambient air.

Needlessly Fast-Turnover Proteins As Next-Gen Synbio Targets

PLANT SYNTHETIC
BIOLOGY, BIOENGINEERING,
& BIOTECHNOLOGY

- Conference: International Conference on Plant Synthetic Biology and Bioengineering
- Year: 2019
- Proceeding: 3rd International Conference on Plant Synthetic Biology, Bioengineering, and Biotechnology
- Group: General Submissions
- Session:
  Engineering Primary Productivity
- Time:
  Sunday, October 6, 2019 - 10:00am-10:30am

Authors:
Hanson, A. D., University of Florida
Joshi, J., Western University
Reisch, C. R., University of Florida

Carbon supply often limits or co-limits crop productivity. A quarter of carbon fixed by photosynthesis is lost via the respiration needed to supply the energy demands of protein turnover and other maintenance processes. Decreasing the energy needed by maintenance processes could therefore increase yield by making more carbon and energy available. Until now, SynBio has been used to increase yield via carbon gain (photosynthesis), not carbon loss, in part because molecular targets to reduce carbon loss have been hard to define. Proteomics now enables identification of 'carbon loss' targets through measuring individual protein turnover rates. One such target is THI4, the enzyme that forms the thiazole moiety of thiamin. THI4 turns over exceptionally fast, and bioenergetic calculations show this turnover could account for 5% of maintenance costs, or 1% of carbon fixed by photosynthesis. Can we convert THI4 from a high-turnover energy hog to a low-turnover, energy-efficient enzyme? Maybe yes. Plant THI4s turn over fast because they are suicide enzymes that use an essential active-site cysteine residue as sulfur donor for the reaction and are inactivated after just one reaction cycle. In contrast, some THI4s from anaerobic prokaryotes use sulfide as sulfur donor and are energy-efficient catalysts; these enzymes have no active-site cysteine. Screening such THI4s for protein expression and functional complementation of an *Escherichia coli* thiazole auxotroph identified some that can function in aerobic conditions like those in plant cells. However, the low sulfide levels *in planta* may limit the activity of such sulfide-utilizing THI4s. We are exploring directed evolution with the EvolvR and OrthoRep systems to produce THI4s that function well in aerobic, low-sulfide conditions. Replacing a suicidal THI4 with an evolved energy-efficient one could slash the energy cost of THI4 turnover by >99.9% and save enough energy to boost biomass yield by 3-4%.

A Synthetic Light-Gated Gene Expression Switch for Plants

PLANT SYNTHETIC
BIOLOGY, BIOENGINEERING,
& BIOTECHNOLOGY

- Conference: International Conference on Plant Synthetic Biology and Bioengineering
- Year: 2019
- Proceeding: 3rd International Conference on Plant Synthetic Biology, Bioengineering, and Biotechnology
- Group: Poster Session
- Session:
  Poster Session
- Time:
  Friday, October 4, 2019 - 5:00pm-6:00pm

Authors:
Larsen, B., The Sainsbury Laboratory Cambridge University
Jones, A., The Sainsbury Laboratory Cambridge University

In the process of applying optogenetic sensors in plants to study phytohormone patterning it has become clear that high-resolution sensing alone is not sufficient - we also need to be able to perturb phytohormone patterning in high-resolution to distinguish correlation from causation. Optogenetic actuators are minimally invasive genetically encoded tools for modifying cell processes with maximal spatiotemporal resolution, but a major challenge for designing optogenetic actuators for plants is that they need to exhibit one target gene expression state during plant growth in normal light-dark cycles, and an altered target gene expression state following treatment with light spectra that are not found in a horticultural environment. We have created an orthogonal synthetic light-gated gene expression system for plants, by repurposing the cyanobacterial CcaS-CcaR system, a photoreversible two-component signal transduction systems, by semi-rational engineering. When tested by transiently transfecting N. benthamiana, our system was inactive under blue light conditions and active in the dark and under green, red and white light conditions. Furthermore, the systems activity could be tuned by modulating the red-to-blue ratio of white light. We are presently engineering the system for function in stably transformed Arabidopsis plants and towards being inactive in ambient white light and in the dark, be activated by green light and inactivated by blue or red light. With these properties, the system will enable us to manipulate biological processes with high-spatiotemporal resolution in a minimally invasive manner.

Direct Detection and Visualization of UV-Induced DNA Damage

PLANT SYNTHETIC BIOLOGY, BIOENGINEERING, & BIOTECHNOLOGY

- Conference: International Conference on Plant Synthetic Biology and Bioengineering
- Year: 2019
- Proceeding: 3rd International Conference on Plant Synthetic Biology, Bioengineering, and Biotechnology
- Group: Poster Session
- Session:
  Poster Session
- Time:
  Friday, October 4, 2019 - 5:00pm-6:00pm

Authors:
Carter, K., Northwestern University

As a result of various ozone-depleting substances progressively damaging our planet's atmosphere, UV exposure from sunlight is at an all-time high, resulting in an increased incidence of skin cancer in many populated areas. This is a pressing concern for areas with high UV exposure, including Peru, Argentina, New Zealand, and Australia with the latter two countries having the highest skin cancer rates in the world. As such, there is a dire need for an easily accessible method for civilians to determine the amount of UV exposure in their area.

Standard UV detection methods currently available such as HPLC and digital UV meters do not meet the need because they involve detection instruments that are nonportable and/or expensive, making them not only impractical for everyday use by civilians but also unavailable to those in resource-limited settings. Here, we leveraged synthetic biology tools to try and create a whole-cell biosensor that can detect UV light rapidly and at a low-cost.

To achieve this goal, we took advantage of the SOS system in *E.coli* competent cells. The SOS system responds to UV-induced DNA lesions by halting the cell cycle and activating the nucleotide excision repair (NER) mechanism. The NER mechanism in *Escherichia coli* begins with the protein UvrA, whose expression increases by 10-fold from around 25 molecules to 250 molecules upon SOS induction. We sought to measure this increase via a plasmid with GFP downstream of the UvrA promoter. This be easily visualized with a low-cost, 3D printable hand-held illuminator, similar to the ones that have been recently developed by the Lucks lab. By quantifying the level of fluorescence, we will be able to measure the UV exposure to the cell. This biological system will be integrated into a portable biosensor that can be used outside of the laboratory setting.

Evaluation of Different Carbon Sources for the Production of Single Cell Protein from Saccharomyces Cerevisiae

PLANT SYNTHETIC BIOLOGY, BIOENGINEERING, & BIOTECHNOLOGY

- Conference: International Conference on Plant Synthetic Biology and Bioengineering
- Year: 2019
- Proceeding: 3rd International Conference on Plant Synthetic Biology, Bioengineering, and Biotechnology
- Group: Poster Session
- Session:
  Poster Session
- Time:
  Friday, October 4, 2019 - 5:00pm-6:00pm

Authors:
Iqbal, M., Al.Moiz industries

Among the prevalent Single Cell Protein sources yeast based SCP supplements are becoming popular due to multiple benefits.This study was plannee for the production of single cell protein through yeast fermentation and evaluation of its nutritive value by proximate analysis. For this purpose Saccharomyces cerevisiae was used. Ten different media were analysed for the production of SCP. Out of ten, one media YPD was used as standard media for the cell biomass production and other media were compared with the standard media. First screening of six media was done in test tubes and further screening of media was done in 250ml flasks and six litter flask. After screening of ten media at different stages only one media YCM ,which is cheap and easily available gave maximum yield of dry Cell biomass comparable with the standard media was compared at fermentor level with the standard media. Other carbon soysour like glucose, Sucrose, maltose, different crop residues like mungbean, black grames, moringa leaf meal, and various concentrations of corn steeping liquor was also used for growth optimization of Saccharomyces cerevisiae. YCM media gave maximum yield of cell biomass as compared to other media used in my research work.

Fluorescence-Based Whole Plant Imaging and Phenomics

PLANT SYNTHETIC
BIOLOGY, BIOENGINEERING,
& BIOTECHNOLOGY

- Conference: International Conference on Plant Synthetic Biology and Bioengineering
- Year: 2019
- Proceeding: 3rd International Conference on Plant Synthetic Biology, Bioengineering, and Biotechnology
- Group: Poster Session
- Session:
  Poster Session
- Time:
  Friday, October 4, 2019 - 5:00pm-6:00pm

Authors:
Rigoulot, S. B., University of Tennesse
Stewart, N., Department of Plant Sciences
Lenaghan, S., University of Tennessee
Schimel, T., Center for Agricultural Synthetic Biology
Meier, K. A., Center for Agricultural Synthetic Biology
Lee, J., The University of Tennessee, Knoxville
Occhialini, A., Center for Agricultural Synthetic Biology
Madaijan, J., Special Technologies Laboratory
Schmid, M., Center for Agricultural Synthetic Biology
DiBenedetto, J., Special Technologies Laboratory
Brabazon, H., Center for Agricultural Synthetic Biology
Seaberry, E. M., Center for Agricultural Synthetic Biology
Layton, J., Center for Agricultural Synthetic Biology
Brabazon, J., Center for Agricultural Synthetic Biology

Reverse genetics approaches have revolutionized plant biology and agriculture. Phenomics has the prospects of bridging plant phenotypes with genes, including transgenes, to literally transform agricultural fields. Genetically-encoded fluorescent proteins are powerful tools to study gene expression, protein trafficking and plant physiology. While the first instance of canopy imaging of green fluorescent protein (GFP) expression in plants was performed over 20 years ago, modern phenomics has largely ignored fluorescence as a transgene indicator in spite of the tremendous color palette of fluorescent proteins available to plant biologists . Here we show a new platform for standoff imaging whole plants expressing a wide variety of fluorescent proteins (FPs) in leaves. The platform, the fluorescence-inducing laser projector (FILP), uses compact diode lasers of various colors and emission filters to phenotype transgenic plants expressing multiple FPs. Of over 20 FPs screened, we selected top performing variants for standoff phenomics at 3 m using a laboratory-based laser range. While FILP is an unprecedentedly versatile laboratory platform, we envisage making minor modifications for automated greenhouse use and even drone-fielded versions of the platform.

Heterologous Production of Forskolin in the Cyanobacterium Synechocystis PCC 6803

PLANT SYNTHETIC BIOLOGY, BIOENGINEERING, & BIOTECHNOLOGY

- Conference: International Conference on Plant Synthetic Biology and Bioengineering
- Year: 2019
- Proceeding: 3rd International Conference on Plant Synthetic Biology, Bioengineering, and Biotechnology
- Group: Poster Session
- Session:
  Poster Session
- Time:
  Friday, October 4, 2019 - 5:00pm-6:00pm

Authors:
Chuk Sutardja, L., University of Copenhagen
Jensen, P. E., University of Copenhagen
Møller, B. L., University of Copenhagen
Nielsen, A. Z., University of Copenhagen
Gnanasekaran, T., University of Copenhagen
Walby, S. L., University of Copenhagen

Photosynthetic microbes, including cyanobacteria, have the remarkable capability to use light as their main source of energy for the production of biomass. In the past decade, there has been an increasing amount of focus put on the sustainable production of high value compounds in cyanobacteria. Efforts include the production of terpenoids, which are used as flavorants, fragrances, and pharmaceuticals. With the expression of various terpene synthases and auxiliary enzymes, it is possible to produce a wide range of terpenoids. Attempts at production are often hampered by low titers, indicating insufficient enzyme levels and low flux through desired pathways. Our work demonstrates the production of the diterpenoid cAMP booster forskolin in *Synechocystis PCC. 6803*. The full biosynthetic pathway consisting of six enzymes was heterologously expressed resulting in a strain capable of producing forskolin. Strains producing forskolin did not show any demonstrable growth defects. Forskolin itself was found secreted in the media, which could potentially allow for continuous harvesting of the compound, akin to a synthetic source leaf. We hypothesized that supply of the precursor geranylgeranyl-diphosphate (GGDP) was posing a bottleneck. Through the overexpression of key enzymes in the MEP pathway, we show a 15-fold improvement of the precursor 13R-manoyl oxide when compared to the naïve expression of the terpene synthases alone.

High-Efficiency, Heritable Gene Editing Using RNA Viruses

PLANT SYNTHETIC
BIOLOGY, BIOENGINEERING,
& BIOTECHNOLOGY

- Conference: International Conference on Plant Synthetic Biology and Bioengineering
- Year: 2019
- Proceeding: 3rd International Conference on Plant Synthetic Biology, Bioengineering, and Biotechnology
- Group: Poster Session
- Session:
  Poster Session
- Time:
  Friday, October 4, 2019 - 5:00pm-6:00pm

Authors:
Ellison, E., University of Minnesota

Gene editing with sequence specific nucleases offers great promise for advancing basic and applied plant research. Nearly all heritable mutations created from gene editing reagents, however, rely on tissue-culture to regenerate edited plant cells. Tissue-culture works on limited genotypes, is very time consuming, and is often technically challenging which dramatically limits the scope and throughput of gene editing in plants. One potential solution is through the use of RNA viral vectors to deliver CRISPR reagents. Up until this point this approach has been limited by low frequency somatic gene editing and the inability to efficiently generate heritable indels. Through the use of augmented sgRNAs expressed from RNA Viral Vectors we have demonstrated high frequency somatic and heritable tissue-culture free gene editing in *Nicotiana benthamiana.* We anticipate this method will be broadly applicable and considerably increase the feasibility of plant gene editing.

Improving Nicotiana Benthamiana As Bioproduction System for Proteins and Small Molecules

PLANT SYNTHETIC BIOLOGY, BIOENGINEERING, & BIOTECHNOLOGY

- Conference: International Conference on Plant Synthetic Biology and Bioengineering
- Year: 2019
- Proceeding: 3rd International Conference on Plant Synthetic Biology, Bioengineering, and Biotechnology
- Group: Poster Session
- Session:
  Poster Session
- Time:
  Friday, October 4, 2019 - 5:00pm-6:00pm

Authors:
Dudley, Q., Earlham Institute
O'Connor, S., Max Planck Institute for Chemical Ecology
Patron, N., Earlham Institute

The wild tobacco relative *Nicotiana benthamiana* is a commonly used plant for manufacturing proteins and reconstituting metabolic pathways which produce complex metabolites such as fragrances or medicines. In particular, the plant can transiently express heterologous multi-enzyme pathways in just a few days. However, small molecule compounds and their intermediate pathway metabolites produced in *N. benthamiana* are often over-glycosylated, oxidized/reduced, acylated or modified with glutathione by enzymes native to *N. benthamiana*. This unwanted activity has challenged efforts to reconstitute the 11-step pathway to strictosidine (an alkaloid precursor to the anti-cancer drug vinblastine). Therefore, to improve *N. benthamiana* as a bioproduction platform, we are using the RNA-guided Cas9 nuclease to deactivate the native enzymes that likely make unwanted modifications to the target metabolite. As a proof-of-concept, we have demonstrated activity of SpCas9 nuclease to make targeted double-strand breaks (producing insertion/deletions) that inactivate the peroxidase NbPOX1. Next, we have used transcriptomic and phylogenetic analysis to select candidate genes for inactivation from among hundreds of possible enzymes. Finally, we have adapted cell-free protein synthesis (CFPS) to manufacture candidate glycosyltransferases to screen their activity towards pathway intermediates geraniol and cis-trans-nepetalactol.

Morphology-Dependent Biomolecule Delivery in Plants with Gold Nanoparticles

PLANT SYNTHETIC
BIOLOGY, BIOENGINEERING,
& BIOTECHNOLOGY

- Conference: International Conference on Plant Synthetic Biology and Bioengineering
- Year: 2019
- Proceeding: 3rd International Conference on Plant Synthetic Biology, Bioengineering, and Biotechnology
- Group: Poster Session
- Session:
  Poster Session
- Time:
  Friday, October 4, 2019 - 5:00pm-6:00pm

Authors:
Goh, N., University of California
Zhang, H., University of California
Butrus, S., University of Michigan
Landry, M., Chan Zuckerberg Biohub

Plant genetic engineering has the potential to ameliorate the effects of climate change and rising populations via the creation of high-yield crop varieties and climate-resilient plants. To this end, we require a plant genetic engineering toolset that is (i) plant-species independent, and (ii) able to bypass physical barriers present in mature plant tissues. A principal challenge remains the efficient delivery of biological cargo across a plant's cell wall and cell membrane. Traditional plant genetic engineering tools such as *Agrobacterium* and biolistic delivery suffer limitations in species range, cargo diversity, tissue damage, efficiency, and transgene integration resulting in regulatory oversight. Recently, nanoparticles (NPs) have emerged as promising materials for use as cargo carriers into plant cells showing high transgene expression without transgene integration.

A lack of heuristics in NP design for unassisted (non-biolistic) DNA, RNA, and protein delivery in plants limits a rational design-based approach for future developments in plant biomolecule delivery. To this end, we utilize DNA-coated gold nanoparticles (AuNPs) as model materials to elucidate the effect of NP morphology on internalization timescale, efficiency, and mode of uptake. We investigate the impact of NP morphology for fluorescent NP internalization into transgenic GFP *Nicotiana benthamiana* (GFP *Nb*) plant cells, demonstrating that size and shape are both internalization-determining parameters. By comparing NP fluorescence against cytosolic GFP fluorescence, we show nanorods experience peak internalization faster than nanospheres of a similar diameter. We then investigate endocytosis as a mechanism of NP uptake through the use of an endocytosis inhibitor, demonstrating morphology-dependent effects. Finally, we build upon morphology studies to deliver siRNA-loaded AuNPs to GFP *Nb* and demonstrate unassisted NP-based gene silencing in mature plants. Our study confirms that NP morphology impacts both time required for and extent of internalization, and confirms NP utility for biomolecule delivery in plants.

New Tools for Engineering the Arabidopsis Plastid Genome

PLANT SYNTHETIC BIOLOGY, BIOENGINEERING, & BIOTECHNOLOGY

- Conference: International Conference on Plant Synthetic Biology and Bioengineering
- Year: 2019
- Proceeding: 3rd International Conference on Plant Synthetic Biology, Bioengineering, and Biotechnology
- Group: Poster Session
- Session:
  Poster Session
- Time:
  Friday, October 4, 2019 - 5:00pm-6:00pm

Authors:
LaManna, L., Rutgers University
Yu, Q., Rutgers University
Kelly, M., Rutgers University
Lutz, K., Farmingdale State College
Maliga, P., Rutgers University

Plastid transformation in the spectinomycin-hypersensitive Col-0 *acc2*-1 and Sav-0 backgrounds was found to be 100-fold more efficient than the wild type (Yu et al., 2017). As a follow up to this study, we developed new tools to expand plastid transformation biotechnology in Arabidopsis. Newly constructed pMEK vectors harbor a left targeting region truncated at the *Hind*III site in the plastid *rrn16* gene to enable utilization of the *Eco*RI-*Hind*III multiple cloning site. A one-step selective SPEED medium permits rapid evaluation of plastid transformation efficiency in 35 to 42 days (5 to 6 weeks) and transplastomic cells can be readily identified by GFP fluorescence. Mutations in the *ACC2* N-terminal chloroplast transit peptide of RLD and Ws produced *ACC2*-knockout lines that are spectinomycin-hypersensitive and pliable for plastid transformation. We have advanced efforts to develop plastid transformation in Arabidopsis by developing a system for rapid evaluation of plastid transformation efficacy, complete with new vectors, culture media, and new *ACC2*-knockout lines that expand the range of accessions available for experimentation.

Novel Associations between Phosphate Accumulating Fungi and Plants for Improved P Uptake

PLANT SYNTHETIC
BIOLOGY, BIOENGINEERING,
& BIOTECHNOLOGY

- Conference: International Conference on Plant Synthetic Biology and Bioengineering
- Year: 2019
- Proceeding: 3rd International Conference on Plant Synthetic Biology, Bioengineering, and Biotechnology
- Group: Poster Session
- Session:
  Poster Session
- Time:
  Friday, October 4, 2019 - 5:00pm-6:00pm

Authors:
van Lierop, L., University of Minnesota

Phosphorus (P), an essential macronutrient for plant growth, is vital for several key plant functions including energy transfer, photosynthesis, transformation of sugars and starches, nutrient movement within plants, and growth and development of new tissues. Although the total P content in soil (average 0.05% w/w) is sufficient for plant growth, only 0.1% of the total P in soil is plant available phosphate. Phosphorus uptake by plants from soil can be greatly improved by fungal root symbionts such as arbuscular mycorrhizal fungi (AMF) and phosphate solubilizing organisms (PSO). This research investigates the possibility of phosphate accumulating organisms (PAO), which are capable of rapid and massive accumulation of polyphosphate in microorganisms, to not only function similarly to mycorrhizal fungi in aiding P acquisition but also facilitate the transformation of organic P compounds from soil for more efficient utilization of plant unavailable P. The results of this study also have potential industrial applications for biofertilizer development to support global food production.

Plant Synthetic Biology: Design, Quantification and Amplification of a Cell Type Specific Synthetic Genetic Circuit in Arabidopsis

PLANT SYNTHETIC
BIOLOGY, BIOENGINEERING,
& BIOTECHNOLOGY

- Conference: International Conference on Plant Synthetic Biology and Bioengineering
- Year: 2019
- Proceeding: 3rd International Conference on Plant Synthetic Biology, Bioengineering, and Biotechnology
- Group: Poster Session
- Session:
  Poster Session
- Time:
  Friday, October 4, 2019 - 5:00pm-6:00pm

Authors:
Oehmke, S., Colorado State University
Medford, J. I., Colorado State University

Synthetic biologists can utilize plants as platforms to create photosynthetically driven and self-sustainable genetic circuits. Furthermore, there are many downstream applications for rationally designed plants. Although plants are excellent platforms, issues and unpredictability arise from the innate complexity of multicellularity and its underlying genetic regulation. The ability to quantitatively control gene expression within specific cell types can address some issues arising from multicellularity. In our research, we have developed a genetic circuit with the ability to induce and quantitatively control expression in *Arabidopsis thaliana* root epidermal cells. Our circuit design uses an externally applied ligand that activates a computationally designed transcriptional response driven by a tissue specific promoter to control output (GFP expression). In addition, we engineered a circuit that adds a positive feedback motif to amplify the system. To quantify the behaviors of these circuits we developed a Matlab program to remove autofluorescence from optical sections captured on a spinning disk confocal micrscope. The program subsequently defines a region of interest, and computes the mean log pixel intensity of the GFP signal within the region. We will present results describing our data to date and on-going work showing this genetic circuit is highly specific for the cell type (root epidermis), fully controllable with the external ligand, and capable of amplification with the positive feedback circuit. The concepts and components of these circuits can be implemented in future designs to engineer and produce plants with more predictable behavior affording the operator greater control.

Accelerated Plant Breeding By Nanomaterial-Delivered Gene-Editing

PLANT SYNTHETIC
BIOLOGY, BIOENGINEERING,
& BIOTECHNOLOGY

- Conference: International Conference on Plant Synthetic Biology and Bioengineering
- Year: 2019
- Proceeding: 3rd International Conference on Plant Synthetic Biology, Bioengineering, and Biotechnology
- Group: Poster Session
- Session:
  Poster Session
- Time:
  Friday, October 4, 2019 - 5:00pm-6:00pm

Authors:
Gonzalez-Grandio, E., UC Berkeley
Demirer, G. S., University of California
Landry, M., Chan Zuckerberg Biohub

A primary focus of the rapidly growing field of plant synthetic biology is to develop technologies to precisely regulate gene expression and engineer complex genetic circuits into plant chassis. At present, there are few orthogonal tools available for effectively controlling gene expression in plants, with most researchers instead using a limited set of viral elements or truncated native promoters. A powerful repressible—and engineerable—binary system that has been repurposed in a variety of eukaryotic systems is the Q-system from *Neurospora crassa*. Here, we demonstrate functionality of the Q-system in plants through transient expression in potato (*Solanum tuberosum*) protoplasts and agroinfiltration in *Nicotiana benthamiana* leaves. Further, using functional variants of the QF transcriptional activator, it was possible to tune the expression of reporter genes. In addition, complete suppression of the system by the QS suppressor was achieved, along with quinic acid-induced reactivation. As a potential application for plant biosensors, we demonstrated the ability of the Q-system to amplify the signal from a weak promoter, enabling remote detection of a fluorescent reporter that was previously undetectable. Based on the results from this study, the Q-system represents a powerful orthogonal tool for precise control of gene expression in plants, with envisioned applications in metabolic engineering, plant biosensors, and biotic and abiotic stress tolerance.

Prediction of Pesticides Residues Quantities in a Proposed Food and Suggestion of Suitable Utilization for Pesticides in a Long-Term Usage

PLANT SYNTHETIC
BIOLOGY, BIOENGINEERING,
& BIOTECHNOLOGY

- Conference: International Conference on Plant Synthetic Biology and Bioengineering
- Year: 2019
- Proceeding: 3rd International Conference on Plant Synthetic Biology, Bioengineering, and Biotechnology
- Group: Poster Session
- Session:
  Poster Session
- Time:
  Friday, October 4, 2019 - 5:00pm-6:00pm

Authors:
Noureldien, M., University of Khartoum

Pesticides are defined as a substance or a mixture of substances used for preventing, destroying, repelling, or mitigation of any pest. They are Usually classified by the intended use, level of toxicity however this article is concerned with the classification by the type of the pest they control. It has been proved that from the millions of tons of the utilized pesticide only 1% reach the targeted pest. The rest appear as residues in food, soil, water in addition to air and often linked these residues with long-term side effects including neurological problems, bird defects, neurodevelopmental disorder, environmental impact on animals, cancer and weather. Despite the fact that most of the pesticides have a relative short effect on the targeted region and the targeted pests itself, only the chlorinated hydrocarbons (e.g. chlordane, Aldrin, DDT, HCH), in addition to limited organophosphorus pesticides (e.g. carbophenothion) have a known long residual action. On the other hand, it is difficult to predict whether a proposed new pesticide can or can't cause these impacts since they are long term side effects. This article propose a numerical method termed as *(Numerical Oriented Usage Regulator)* to predict whether a proposed pesticide is significantly likely to make hazard residues in selected sort of food and suggest the regulations of the utilization of the proposed Pesticide. The US PDP's (Pesticides Data Program) data from 1992 till 2017 were used as a data source to implement and test the proposed method. The objective of this article to assist the scientists to predict the possibility of a proposed pesticide to make above-limited residues in food which is a long term side effect and hard to predict, by using a relatively short term scale.

Programmable Detection of Target RNA Sequences In Vivo

PLANT SYNTHETIC
BIOLOGY, BIOENGINEERING,
& BIOTECHNOLOGY

- Conference: International Conference on Plant Synthetic Biology and Bioengineering
- Year: 2019
- Proceeding: 3rd International Conference on Plant Synthetic Biology, Bioengineering, and Biotechnology
- Group: Poster Session
- Session:
  Poster Session
- Time:
  Friday, October 4, 2019 - 5:00pm-6:00pm

Authors:
Nou, X. A., MIT
Voigt, C. A., Massachusetts Institute of Technology
Segall-Shapiro, T., Massachusetts Institute of Technology

The mRNA profile of a cell provides critical information, including metabolic state, cellular stress, and foreign gene acquisition. While there are methods to determine what mRNAs are present, they require cell lysis, insertion of potentially disruptive sequences, or massive overexpression of the detector. Here, we propose to detect a target RNA *in vivo* using two orthogonal dCas13s that use gRNAs to bind to two adjacent locations, thus inducing the proximity-dependent release of a prokaryotic sigma factor via intein trans-splicing. To this end, we evaluated the ability for four dCas13s, from *Ruminococcus flavefaciens*, *Prevotella sp. P5-125*, *Porphyromonas gulae*, and *Riemerella anatipestifer,* to target a ribosome binding site (RBS) controlling red fluorescent protein (RFP) in *Escherichia coli* and confirmed RNA binding by measuring a reduction in fluorescence. One gRNA was found to be toxic, suggesting off-target binding to an endogenous mRNA. Separately, we have designed a system where a protein co-localization event causes two domains of a *Pseudomonas fluorescens* sigma factor to covalently fuse via an intein. As a proof-of-principle, a rapamycin-driven protein-protein interaction is used to induce expression of a fluorescent reporter from the sigma factor-specific promoter. Finally, we are in the process of using pairs of dCas13:gRNAs to drive the same proximity-dependent sigma factor release by binding to adjacent sites on a target RNA. Current iterations of this system have not been responsive to the presence of the target mRNA: we are currently trying to identify whether the problem may be caused by steric hindrances that disrupt RNA-binding or sigma factor release. If successful, this system would enable the real-time *in vivo* detection of specific mRNAs without external perturbation, for the purposes of both studying endogenous gene expression and engineering systems to respond to intracellular changes such as stress and pathogen infection.

Prototyping Nitrogenase Expression in Plastids with a Chloroplast Cell-Free System

PLANT SYNTHETIC
BIOLOGY, BIOENGINEERING,
& BIOTECHNOLOGY

- Conference: International Conference on Plant Synthetic Biology and Bioengineering
- Year: 2019
- Proceeding: 3rd International Conference on Plant Synthetic Biology, Bioengineering, and Biotechnology
- Group: Poster Session
- Session:
  Poster Session
- Time:
  Friday, October 4, 2019 - 5:00pm-6:00pm

Authors:
Majer, E., Massachusetts Institute of Technology
Voigt, C. A., Massachusetts Institute of Technology
Jewett, M. C., Northwestern University
Bock, R., Max Planck Institute of Molecular Plant Physiology
Clark, L. G., Northwestern University
Ryu, M. H., MIT
Karcher, D., Max Planck Institute of Molecular Plant Physiology

Plastids are plant subcellular compartments that are attractive targets of plant engineering because they are amenable to transformation, have high protein expression capacity and are maternally inherited. In order to transfer complex functions, like nitrogen fixing capacity, into plastids, gene expression has to be finely tuned. This requires several design iterations, but the time-consuming transformation of plants keeps engineers from executing this process *in vivo*.

In this work, we used a chloroplast cell-free system to prototype synthetic nitrogenase clusters for plastid expression. The cell-free system, purified from tobacco chloroplasts, has the transcription and translation machinery of this organelle. The clusters were restructured by using well-defined parts, codon-optimizing genes and putting them under the control of the T7 RNA polymerase (T7RNAP) which allows the transfer between species. We show that the chloroplast cell-free system is able to express these synthetic clusters composed of 16-18 genes at mRNA level.

In order to correlate expression in the chloroplast cell-free system with *in vivo* data, we transformed these clusters into tobacco chloroplasts along with the T7RNAP, which was put under the control of an inducible riboswitch that allows translation upon theophylline induction. RNA-seq data of these plants showed that T7RNAP responds to the chemical induction and all clusters are expressed at the mRNA level. *In vivo* data obtained from these plants correlated well with our chloroplast cell-free data demonstrating that the chloroplast cell-free system can serve to prototype multi-genic constructs before plant transformation.

Our chloroplast cell-free system could allow the precise characterization of genetic parts and the testing of 10^3-10^4 construct in a week. Those constructs producing the desired expression pattern can be selected for plastid transformation which would reduce the amount of work related to plant transformation and increase the success rate of engineering complex proteins in plants.

The Expression of Japanese Encephalitis Virus Envelope Gene in Green Cos

PLANT SYNTHETIC
BIOLOGY, BIOENGINEERING,
& BIOTECHNOLOGY

- Conference: International Conference on Plant Synthetic Biology and Bioengineering
- Year: 2019
- Proceeding: 3rd International Conference on Plant Synthetic Biology, Bioengineering, and Biotechnology
- Group: Poster Session
- Session:
  Poster Session
- Time:
  Friday, October 4, 2019 - 5:00pm-6:00pm

Authors:
Tangkananond, W., Faculty of Science and Technology, Thammasat University, Rangsit Center, 99 Mo 18 Phahonyothin Road, Klong 1 District, Klong Luang
Lomonossoff, G. P., John Innes Centre

Abstract
The expression of Japanese encephalitis virus (JEV) envelope gene in green cos leaf was a project undertaken at the John Innes Centre (JIC), Norwich, UK during 6 September to 20 October 2018 and is one of the projects of the research training program at the JIC during long vacation every year since 2010 until now.

The summary of this research is as follows. The production of envelope protein of the JEV in green cos by using the envelope gene expression by 8 epitopes (MEP: multiepitope) of Japanese encephalitis virus vaccine strain SA14-14-2 that provides the highest immune response against JEV. After incorporated the envelope gene into pEAQ-HT vector, we cloned genes in *Escherichia coli* and *Agrobacterium tumefaciens*, respectively, then expressed the envelope protein of JEV in green cos leaves. The result can be seen in photographs of green cos leaves by both visible light camera and ultraviolet light camera.

Further research should include the analysis and identification by chemiluminescence immunoassay and by the Matrix-assisted Laser Desorption/Ionization Time of Flight (MALDI-TOF). Further purification of the envelope protein expression of JEV in green cos that will be benefited for the production of envelope protein of the JEV plant-based vaccine in swine.

Key words: *Japanese encephalitis virus*, plant-based vaccine, chemiluminescence immunoassay

The Role of Pollination in Flower Senescence and the Expression of Ethylene Biosynthesis Genes in Tomato (Solanum lycopersicum)

PLANT SYNTHETIC BIOLOGY, BIOENGINEERING, & BIOTECHNOLOGY

- Conference: International Conference on Plant Synthetic Biology and Bioengineering
- Year: 2019
- Proceeding: 3rd International Conference on Plant Synthetic Biology, Bioengineering, and Biotechnology
- Group: Poster Session
- Session:
  Poster Session
- Time:
  Friday, October 4, 2019 - 5:00pm-6:00pm

Authors:
Awasthi, S., University of Cambridge

Flower senescence is the end phase of flower development, leading to the termination of the flower, including flower wilting and the dehiscence of flower organs. Flower senescence and fruit ripening are the most important physiological responses that are controlled by ethylene as a major regulator. These phenomena require significant attention in terms of crop improvement, utilizing molecular genetics as well as conventional breeding approaches. We are exploring whether pollination-driven ethylene production is responsible for flower senescence in tomato (*Solanum lycopersicum*), a model crop plant, using flowers from ethylene insensitive mutants *non-ripening* (*nor*) and *alcobaca* (*alc*). qRT-PCR suggests that expression of ethylene biosynthetic genes is delayed in *nor* and *alc* post-pollination, relative to an ethylene sensitive control line, as a result of post pollination events occurring at different time points. Microscopic analysis of pollen tube growth confirms a positive correlation between the pollination event and *SlACS* (ethylene biosynthetic gene) expression. Our initial results, therefore, suggest that *SlACS* gene expression is crucial for the preliminary spurt of ethylene production which is eventually responsible for flower senescence.

Using Synthetic Biology to Produce High-Value Diterpenoids in Chlamydomonas Reinhardtii

PLANT SYNTHETIC
BIOLOGY, BIOENGINEERING,
& BIOTECHNOLOGY

- Conference: International Conference on Plant Synthetic Biology and Bioengineering
- Year: 2019
- Proceeding: 3rd International Conference on Plant Synthetic Biology, Bioengineering, and Biotechnology
- Group: Poster Session
- Session:
  Poster Session
- Time:
  Friday, October 4, 2019 - 5:00pm-6:00pm

Authors:
Mehrshahi, P., University of Cambridge
Smith, A., University of Cambridge
Gorchs Rovira, A., University of Cambridge

The demand for food, fuel and pharmaceuticals by an ever-growing world population is fast outstripping sustainable supply. Adoption of industrial biotechnology platforms for synthesis of high-value and high-demand products provides a means to improve the sustainable supply bottleneck. While large-scale bacterial and yeast platforms are mature industries, there are biological constraints on the range of molecules that can be synthesised by these organisms. Photosynthetic algal platforms are capable of production of diverse and complex compounds, including vitamins such as carotenoids (pro-vitamin A), oils (such as omega-3 fatty acids), and by genetic manipulation, novel compounds for pharmaceuticals, such as plant natural products or glycosylated antigens. However, despite some success stories, the low yields reported for many target products are a major obstacle to up-scaling and commercial viability of algal platforms.

We have exploited the green microalga, *Chlamydomonas reinhardtii*, as a chassis organism for scalable and high-throughput production of medicinal compounds. We are taking a synthetic biology driven approach, following the principle of design, build, and test to overcome the scalability constraints associated with first-generation engineered algal platforms. Using functionally defined standardised parts and workflows, we are able to predictably generate *C. reinhardtii* strains that efficiently express multiple heterologous genes. For control of transgene expression, we have developed a versatile and easily regulated system based on thiamine-pyrophosphate riboswitches, which are responsive to the concentration of thiamine and its analogues in the growth medium. Using these strategies and molecular resources we are engineering *C. reinhardtii* for scalable production of diterpenoids with medicinal properties.

Scranton, M. A. *et al. Algal Research* **15,** 135–142 (2016)

Scaife, M. A. *et al. The Plant Journal* **82,** 532–546 (2015)

Nguyen, G. T. *et al*. J. Phycology. **52**, 320–328 (2016).

Keyword
Synthetic biology, metabolic engineering, microalgae, riboswitch, diterpenoids

Utilization of the Q-System for Synthetic Transcriptional Regulation in Plants

PLANT SYNTHETIC
BIOLOGY, BIOENGINEERING,
& BIOTECHNOLOGY

- Conference: International Conference on Plant Synthetic Biology and Bioengineering
- Year: 2019
- Proceeding: 3rd International Conference on Plant Synthetic Biology, Bioengineering, and Biotechnology
- Group: Poster Session
- Session:
  Poster Session
- Time:
  Friday, October 4, 2019 - 5:00pm-6:00pm

Authors:
Persad, R., Center for Agricultural Synthetic Biology
Reuter, D., Center for Agricultural Synthetic Biologu
Dice, L. T., Center for Agricultural Synthetic Biology
Nguyen, M. A., Center for Agricultural Synthetic Biology
Occhialini, A., Center for Agricultural Synthetic Biology
Stewart, N., Department of Plant Sciences
Lenaghan, S., University of Tennessee

A primary focus of the rapidly growing field of plant synthetic biology is to develop technologies to precisely regulate gene expression and engineer complex genetic circuits into plant chassis. At present, there are few orthogonal tools available for effectively controlling gene expression in plants, with most researchers instead using a limited set of viral elements or truncated native promoters. A powerful repressible—and engineerable—binary system that has been repurposed in a variety of eukaryotic systems is the Q-system from *Neurospora crassa*. Here, we demonstrate functionality of the Q-system in plants through transient expression in potato (*Solanum tuberosum*) protoplasts and agroinfiltration in *Nicotiana benthamiana* leaves. Further, using functional variants of the QF transcriptional activator, it was possible to tune the expression of reporter genes. In addition, complete suppression of the system by the QS suppressor was achieved, along with quinic acid-induced reactivation. As a potential application for plant biosensors, we demonstrated the ability of the Q-system to amplify the signal from a weak promoter, enabling remote detection of a fluorescent reporter that was previously undetectable. Based on the results from this study, the Q-system represents a powerful orthogonal tool for precise control of gene expression in plants, with envisioned applications in metabolic engineering, plant biosensors, and biotic and abiotic stress tolerance.

An Emerging Technology for Reduction of Chlorella Associated Bacterial Loads for Quality Biomass Production without Affecting Chlorella

PLANT SYNTHETIC
BIOLOGY, BIOENGINEERING,
& BIOTECHNOLOGY

- Conference: International Conference on Plant Synthetic Biology and Bioengineering
- Year: 2019
- Proceeding: 3rd International Conference on Plant Synthetic Biology, Bioengineering, and Biotechnology
- Group: Poster Session
- Session:
  Poster Session
- Time:
  Friday, October 4, 2019 - 5:00pm-6:00pm

Authors:
C K, M. Sr., CSIR-CENTRAL FOOD TECHNOLOGICAL RESEARCH INSTITUTE
Mohammad, S. Sr., CSIR-CENTRAL FOOD TECHNOLOGICAL RESEARCH INSTITUTE
S, S. Jr., CSIR-CENTRAL FOOD TECHNOLOGICAL RESEARCH INSTITUTE
R, S. Sr., CSIR-CENTRAL FOOD TECHNOLOGICAL RESEARCH INSTITUTE
N Mudliar, S. Sr., CSIR-CENTRAL FOOD TECHNOLOGICAL RESEARCH INSTITUTE

Microalgae are symbiotically associated with bacteria during their growth. Microalgal cultivation in photobioreactors is desirable for high value metabolites due to controlled conditions. During algal cultivation in photobioreactors, bacteria are found to be exponentially increased, which cannot be avoided during the algal growth due to symbiosis. Although the bacteria promote *Chlorella* growth by providing supplements (vitamins, hormones), they increase exponentially and hence the bacterial load needs to be reduced to meet desirable US-FDA standards (~10^3 CFU/mL). This study results showed that higher *Chlorella* associated bacterial load was in the range of 10^3-10^{11} CFU/mL during *Chlorella* cultivation in an airlift photobioreactor. Bacterial load reduction concerning time and ozone concentrations were determined. Ozonation had a significant effect on bacterial load reduction, which was in the range of 3 - 6.5 log. The bacterial disinfection kinetics of ozonation showed an overall second-order reaction with a rate constant (K) of 0.93 mg^{-1} min^{-1} L. Further, confocal laser microscopy imaging of *Chlorella* and its associated bacterial cells indicated a selective reduction of associated bacteria without affecting the *Chlorella* cells, which was further confirmed with trypan viability assay. It can be concluded that optimised conditions (ozone concentration, time) will help in the production of high-quality *Chlorella* biomass with reduced bacterial load through ozonation.

Assessing the Toxicological Risk of Chemical Interactions between Titanium Dioxide Nanoparticles and N,N-Diethyl-3-Methylbenzamide (DEET)

PLANT SYNTHETIC BIOLOGY, BIOENGINEERING, & BIOTECHNOLOGY

- Conference: International Conference on Plant Synthetic Biology and Bioengineering
- Year: 2019
- Proceeding: 3rd International Conference on Plant Synthetic Biology, Bioengineering, and Biotechnology
- Group: Poster Session
- Session:
  Poster Session
- Time:
  Friday, October 4, 2019 - 5:00pm-6:00pm

Authors:
Brown, H. G., Oregon State University
Harper, S., Oregon State University
Harper, B., Oregon State University

Sustainable maize production requires combining multiple biotech and native traits that provide insect control, disease resistance and tolerance to herbicides or abiotic stresses. Because trait genes have been randomly inserted at various genomic locations, combining numerous traits is expensive, time-consuming and complex. Here we report the Complex Trait Locus (CTL) for genetic trait stacking. CTL consisted of multiple preselected sites which were positioned within a small, well characterized chromosomal region. Firstly, a site-specific insertion landing pad (SSILP) was targeted to one of the sites using CRISPR-Cas9 technology, producing individual inbred lines capable of receiving trait genes via recombinase-mediated cassette exchange (RMCE). The traits then can be combined within a linkage block using genetic crossing and bred as a single locus. We generated 57 SSILP lines at 4 CTLs across 3 inbreds. The *NPTII* gene in the SSILP expressed consistently and had minimal impact on neighboring endogenous gene expression. Two-gene stacking was readily achieved. The CTL technology is a major step forward in streamlining biotech product testing and deployment. This study also advances our understanding of gene targeting and expression.

Biosynthesis of Bioactive Diterpenoids: Pathway Discovery and Reconstruction

PLANT SYNTHETIC BIOLOGY, BIOENGINEERING, & BIOTECHNOLOGY

- Conference: International Conference on Plant Synthetic Biology and Bioengineering
- Year: 2019
- Proceeding: 3rd International Conference on Plant Synthetic Biology, Bioengineering, and Biotechnology
- Group: Poster Session
- Session:
  Poster Session
- Time:
  Friday, October 4, 2019 - 5:00pm-6:00pm

Authors:
Pateraki, I., University of Copenhagen
Forman, V., University of Copenhagen
Luo, D., University of Copenhagen
Heskes, A., University of Copenhagen
Møller, B. L., University of Copenhagen
Andersen-Ranberg, J., University of Copenhagen

Plants produce more than 11,000 different diterpenoids which display a wide range of biological activities. Several are used directly as therapeutics for the treatment of human diseases, such as the cyclic AMP booster forskolin or the anti-cancer drugs paclitaxel and ingenol mebutate. Others serve as lead compounds in drug discovery and have been the focus of many medicinal chemistry and therapeutic studies. However, most bioactive diterpenoids have been isolated from non-cultivated medicinal plants or from endangered or red listed species. The lack of a reliable and environmentally benign and stable supply chain thus remains a major obstacle when the need arises to develop specific diterpenoids into medicinal drug. Using forskolin and ingenol mebutate as examples, we provide an account on how production of these plant-derived pharmaceuticals may shift from field acquisition to sustainable manufacturing. To achieve this, knowledge of the enzymes involved in the biosynthetic pathway is essential. Through the strategy consisting of metabolite profiling, transcriptome analysis and enzyme characterization, we have managed to elucidate biosynthetic pathways of several bioactive diterpenoids including forskolin, jolkinol C and vitexilactones[1,2,3]. Subsequently, reconstitution of elucidated pathways in microbial hosts like yeast or photosynthetic microalgae offers the sustainable production of targeted plant diterpenoids.

Reference

(1) Luo D. et al., *Proc. Natl. Acad. Sci. U. S. A.* **2016**, 113, E5082.
(2) Pateraki I. et al., *eLife* **2017**, *6*, e23001.
(3) Heskes AM. et al., *Plant J.* **2018**, *93*, 943.

Biotic Association of Osteopontin in Calcium Carbonate Polymer Formation to Alter the Dissolved CO2 Concentration of Ocean

PLANT SYNTHETIC BIOLOGY, BIOENGINEERING, & BIOTECHNOLOGY

- Conference: International Conference on Plant Synthetic Biology and Bioengineering
- Year: 2019
- Proceeding: 3rd International Conference on Plant Synthetic Biology, Bioengineering, and Biotechnology
- Group: Poster Session
- Session:
  Poster Session
- Time:
  Friday, October 4, 2019 - 5:00pm-6:00pm

Authors:
Jakkal, Y., Student

Calcium carbonate (CaCO3) is one of the best CO2 absorbers. But formation of polymeric sheet of it is a thermodynamically unfavorable reaction which needs a presence of catalyst or heat. But this can be possible by the single enzyme found in hen's uterus which can stabilize the environmental imbalance.

The entire process of eggshell formation during the development of the chick embryo is catalyzed by various enzymes, out of which the enzyme 'osteopontin' is the most important. It is an effort to isolate the specific SPP1 gene from the hen's uterus. This enzyme will help in the polymerization of the CaCO3 subunits during the eggshell formation. Bioinformatic analysis and identification of this gene can be obtained from researches already done in this field and the hen genome database.This SPP1 gene has to be expressed in the mammalian cells or yeast for its expression as the protein is in the glycosylated nature, either by plasmid incorporation or integration into the host genome. The recombinant protein (enzyme) has to place under strong promoters along with the enhancer sequence at the 5' UTR (Untranslated region) of the promoter so that its expression is enhanced manifold and huge amount of recombinant protein will be obtained. The overexpressed protein should then be isolated by using various biophysical techniques. Establishing this protein in the cost effective manner in the larger amount on the industrial scale will aid to form the polymer sheets in the semisolid form by inhibiting the catalysis of reaction. However, using of such sheets on the ocean which are the large reservoir of dissolved CO2 will undergo this reaction and lead to the formation of Calcium Bicarbonate and settle down in the ocean.

Chloroplast-Selective Gene Delivery in Mature Plants Using Carbon Nanotube Nanocarriers

PLANT SYNTHETIC
BIOLOGY, BIOENGINEERING,
& BIOTECHNOLOGY

- Conference: International Conference on Plant Synthetic Biology and Bioengineering
- Year: 2019
- Proceeding: 3rd International Conference on Plant Synthetic Biology, Bioengineering, and Biotechnology
- Group: Poster Session
- Session:
  Poster Session
- Time:
  Friday, October 4, 2019 - 5:00pm-6:00pm

Authors:
Lew, T. T. S., Massachusetts Institute of Technology
Strano, M., Massachusetts Institute of Technology
Kwak, S., Seoul National University

Plant genetic engineering is an important tool used in current efforts in crop improvement, pharmaceutical product biosynthesis and sustainable agriculture. However, conventional genetic engineering techniques are primarily limited to the use of biolistic device or PEG-mediated transformation, the efficiency of which is highly species- and tissue-dependent. In this work, we rationally designed chitosan-complexed single-walled carbon nanotubes, utilizing an experimentally-verified mechanism to safely and selectively deliver genetic materials into the chloroplasts of mature plants. This nanoparticle-mediated delivery does not require external chemical or mechanical aid which often leads to significant damage to the plant tissue. We demonstrate chloroplast-targeted transgene delivery and transient expression in mature *Eruca sativa, Nasturtium officinale*, *Nicotiana tabacum* and *Spinacia oleracea* plants and in isolated *Arabidopsis thaliana* mesophyll protoplasts. Our work provides practical advantages over current delivery techniques as a potential species-independent transformation method to benefit plant bioengineering and biological studies.

CRISPR/CAS9 Mediated Knock out of Multiple Trichome-Regulating MYB Transcription Factors in Populus

PLANT SYNTHETIC BIOLOGY, BIOENGINEERING, & BIOTECHNOLOGY

- Conference: International Conference on Plant Synthetic Biology and Bioengineering
- Year: 2019
- Proceeding: 3rd International Conference on Plant Synthetic Biology, Bioengineering, and Biotechnology
- Group: Poster Session
- Session:
  Poster Session
- Time:
  Friday, October 4, 2019 - 5:00pm-6:00pm

Authors:
Bewg, W., University of Georgia
Chen, M., University of Georgia
Harding, S., University of Georgia
Yu, J., University of Georgia
Jenkins, J., HudsonAlpha Institute for Biotechnology
Barry, K., U.S. Department of Energy Joint Genome Institute
Lipzen, A., U.S. Department of Energy Joint Genome Institute
Schmutz, J., HudsonAlpha Institute for Biotechnology
Tsai, C. J., University of Georgia

The *Populus tremula* x *P. alba* (IRNA 717-1B4) PtaMYB186 transcription factor was previously identified as a regulator of trichome initiation during early stages of leaf development. Here, we employ CRISPR/Cas9 to target *PtaMYB186* and its close paralogs, *PtaMYB138* and *PtaMYB38*, for mutagenesis in *Populus tremula* x *P. alba* using a single gRNA targeting a conserved region. We obtained 37 transgenic events, all with a glabrous phenotype. Amplicon-sequencing confirmed successful edits in all 25 randomly selected mutants, with 24 complete knockouts. RNA-Seq analysis of fully expanded mutant leaves identified 244 and 308 genes that were significantly up- and down-regulated, respectively, when compared with controls. Interestingly, two orthologs of *Arabidopsis* trichome-regulating genes were up-regulated and four orthologs of *Arabidopsis* circadian rhythm genes were down-regulated. The data suggest that mutation of trichome regulators normally expressed during early stages of leaf development can have pleiotropic effects on the transcriptomes of more mature leaves. Gas chromatography–mass spectrometry analysis of leaf cuticle wax composition revealed no major compositional differences, but subtle changes were observed between mutant and control plants. Five mutants were subjected to whole genome re-sequencing and preliminary assessments revealed no evidence of off-target mutations or rearrangements by CRISPR/Cas9 in poplar. Finally, the recently completed genome sequencing of *Populus tremula* x *P. alba* revealed copy number variation that may be linked to natural variation in trichome density among *Populus* species. This research demonstrates that a single variant-free gRNA is highly efficient for multi-allele knockout in poplar. The striking glabrous phenotype of the mutants also highlights the potential of targeting trichome-regulators as a non-lethal visual marker system for assessing new CRISPR reagents or mutagenesis strategies.

Determination of Chemical and Biophysical Properties of Gelatin from Cuttle FISH

- Conference: International Conference on Plant Synthetic Biology and Bioengineering
- Year: 2019
- Proceeding: 3rd International Conference on Plant Synthetic Biology, Bioengineering, and Biotechnology
- Group: Poster Session
- Session:
  Poster Session
- Time:
  Friday, October 4, 2019 - 5:00pm-6:00pm

Authors:
Paramasivam, G. Sr., AAMCE

Gelatin is one of the most popular biopolymers. It is widely used in food, pharmaceutical, cosmetic and photographic application because of its unique functional and technological properties these properties increases the demand of gelatin production in in industries. Fish sausage was prepared from fish by-products, without or with addition of gelatin at different levels. The addition of gelatin to fish sausage resulted in significant improvement of the physiochemical properties. further, the addition of gelatin increased protein content and the moisture and reduced cooking loos, indeed, addition of gelatin yielded products with enhanced textural parameters and the most appreciated sausage was with 1% gelatin .fish sausages can be conserved for 15 days without alterations supplement with the gelatin. The use of bio-based polymer in food packaging industry has recently come to the forefront as a result of the recent crisis awareness with respect to the depletion of non renewable resources. Gelatin production from alternative non mammalian species also gain importance, largely as a way to valorise by products from industrial fisheries processes. Regarding fish origin, cephalopods from tropical and sub tropical waters shown high rheological properties and thermo stability on par with mammal gelatins; therefore, they could be used for similar applications. In this study, gelatin was synthesized from *speiida officinails* popular cuttlefish species of east coast of India the gelatin potentially serve as a promising bio-based packaging material with the application ranging from food to pharmaceutical industries. This study will pave a way for future research in food packaging and meat processing industry.

AUTHOR INDEX